Sustainable Business Practices for Rural Development

Hardeep Chahal • Vijay Pereira
Jeevan Jyoti
Editors

Sustainable Business Practices for Rural Development

The Role of Intellectual Capital

palgrave
macmillan

Editors
Hardeep Chahal
Department of Commerce
University of Jammu
Jammu
Jammu and Kashmir, India

Vijay Pereira
Department of Humanities
and Social Sciences
Khalifa University
Abu Dhabi, United Arab Emirates

Jeevan Jyoti
Department of Commerce
University of Jammu
Jammu
Jammu and Kashmir, India

ISBN 978-981-13-9297-9   ISBN 978-981-13-9298-6 (eBook)
https://doi.org/10.1007/978-981-13-9298-6

Cover illustration: eStudio Calamar

This Palgrave Macmillan imprint is published by the registered company Springer Nature Singapore Pte Ltd.
The registered company address is: 152 Beach Road, #21-01/04 Gateway East, Singapore 189721, Singapore

Foreword

The events of the last two decades have demonstrated that development will not come through abstract trickle-down processes; just because a few individuals or nations are prosperous, does not mean that many will be so. Rather, countries that have made the most impressive progress are those that have provided the support—and space—for individuals to develop their human and social capital and, indeed, physical capital. Individuals do not possess the resultant composite intellectual capital as an abstract commodity; it is something that only exists within a social and economic setting, and is reinforced, developed and disseminated within groupings, communities and societies through a careful mix of regulation and delegation. Indeed, all intellectual capital has individual and collectivist dimensions; it is something that is promoted and sustained through the workings of state, business and society.

Again, talent is often held to be something that is possessed by a few. However, many very poor people exhibit their talents in a wide range of manners, be it in starting enterprises, in undertaking community works, or in taking the decision to reinvent themselves in a different locale. When countries effectively manage talent, it is not simply a product of elites reinvesting in themselves, but rather the recognition that talent takes many forms, and that in the rural periphery, effective management of talent is firstly about according people the opportunity to escape concentrating on daily subsistence and the avoidance of endemic health risks. Given this space, many will rise to the opportunity to better their lives; active developmental policies can monitor the shape such bottom-up betterment efforts take, and redirect resources accordingly.

This initiative by Chahal, Pereira and Gupta brings together accounts from South Asia, and sheds unique light on the intellectual capital-development nexus. It is unique in its juxtaposition and synthesis of relevant parallel bodies of theorising on human and economic development; chapters present fresh evidence on a range of salient initiatives and effectively draw out the implications for future scholarship and policy. As such, it will be of interest to academics and students in a wide range of cognate fields, and to those involved in government, business, NGOs, and at community level in the region and beyond. Indeed, given that many of the issues discussed in the various chapters touch on universal developmental questions, and grapple with solutions that often have much wider applicability, this volume can be considered a genuinely international one; it will be interesting to see how development scholars elsewhere in the world respond to this challenge.

Geoffrey Wood

Dean of Essex Business School
University of Essex
Colchester, UK

Professor of International Business
University of Essex
Colchester, UK

Co-editor in Chief, British Journal of Management
Western University
London, Canada

Contents

Notes on Contributors

Deepti Abrol, PhD is a lecturer in the Department of Commerce, P.G. College for Women, Jammu, India. She has teaching and research experience of seven years. She received her Doctoral Research Degree on "Customer Relationship Management in Insurance Sector: A Study of General Insurance Companies" from the University of Jammu. She has attended various national and international workshops and has participated as well as presented research papers in many national and international conferences. Her research is acknowledged in Journals such as *International Journal of Customer Relations.*

Madhu Bala is Assistant Professor of Commerce at the Government Degree College Kishtwar, Kishtwar, Jammu and Kashmir, India. She obtained her PhD from University of Jammu on the topic "Service brand equity in healthcare sector of India". Her works have been published in journals like *International Journal of Healthcare and Quality Assurance* (Emerald) and *Vision: The Journal of Business Perspective* (Sage).

Hardeep Chahal, PhD is a professor in the Department of Commerce, University of Jammu, India. Her research interests focus on services marketing with emphasis on consumer satisfaction and loyalty, service quality, brand equity and market orientation. Her research work has been acknowledged in refereed international journals and national journals of international repute. She has also co-edited six books including *Research Methodology in Commerce and Management* and *Strategic Service Marketing*. She serves on the editorial boards of the *International Journal of Health Quality and Assurance (Emerald)* and *Journal of Services*

Research (IIMT, India). Overall, she has published more than 100 research papers in journals and books including papers listed in Scopus and ABDC ranked journals. She received the Award of Excellence for Best Reviewer for the year 2011 by *International Journal of Health Quality and Assurance* (Emerald). She has been a visiting fellow at the University of Loughborough, UK under Commonwealth Fellowship Scheme (British Academy Award) and also at the Gandhi Institute of Business and Technology, Jakarta, Indonesia. She has vast administrative experience and is Deputy Coordinator, SAP (Special Assistance Programme, UGC). She is also Rector at Udhampur Campus, which is an offsite campus of University of Jammu.

Kamani Dutta, PhD has teaching and research experience of five years. She received her Doctoral Research Degree on "Exploring determinants, dimensions and outcomes of customer experience in selected services" from University of Jammu. Her research is acknowledged in national and international journals.

Jeevan Jyoti, PhD is an assistant professor at the Department of Commerce, University of Jammu. Her areas of interest are Strategic Human Resource Management, Organisational Behaviour and Entrepreneurship. She has published research papers in prominent national and international journals

Jhanvi Khanna is school teacher in the Jammu and Kashmir State Board of School Education, Jammu, India. She did B.A (Economics Hons.) from Punjab University. Apart from her interest in Economics, she has done her Master's in Education and has been awarded Junior Research Fellowship in Education. She has five years of teaching experience. She has presented papers in various international and national conferences and has attended several capacity-building workshops. She has also authored papers on various economic issues in reputed journals. Her interest lies in the study of population and various socio-economic problems related to it. She has a written record of her knowledge related to demography and women entrepreneurship in the form of a book.

Sumeet Kour, PhD is an assistant professor in Cluster University of Jammu (Jammu & Kashmir), India. Her areas of interest are Entrepreneurship and Human Resource Management. She has received Junior Research Fellowship and Senior Research Fellowship from the UGC. She received Prof. Manubhai M. Shah Memorial Research Award and the Gold Medal for paper titled "Role of Well-being and Mentoring between Perceived High Performance Work System and Intention to Leave: A Structural Model" in the 71st All India Commerce Conference. She received her M. Phil. and PhD from the Department of Commerce, University of Jammu. Kour has written 17 articles in reputed national and international refereed journals including 4 in ABDC, Thomson Reutor ranked Journals.

Pradeep Kumar was a lecturer in the Department of Commerce, Udhampur Campus, University of Jammu and at he works in Government Degree College for Boys, Udhampur (J&K). He has seven years of teaching experience and has attended and presented number of papers in various national and international conferences. He has also written papers in various reputed journals.

Neetu Kumari, PhD has rich experience in teaching and research and is a lecturer in the Department of Commerce, Udhampur Campus, University of Jammu, India. Her research interests focus on Service Marketing and Human Resource Management with emphasis on consumer satisfaction, loyalty, service quality, and perceived value. Her research work has been acknowledged in refereed international journals. She has authored a book on E-commerce which was published in 2015 and enjoyed tremendous success and was highly appreciated by academicians, scholars and students. She is a frequent speaker in national and international conferences on several topics related to marketing, human resource and entrepreneurship. She has authored more than 20 research papers, case studies and chapters.

Sushmera Manikandan is a research scholar at Swiss Business School, Switzerland. She holds a Master's Degree from the University of Wollongong in Dubai and a Master's Degree from Anna University, India. Manikandan has worked as Adjunct faculty at IMT Dubai and Manipal University Dubai. She has 22 research papers to her credit, which were published in logistics and operations management journals.

Sandeep Narula, PhD has 30 years of industry and academic exposure and has authored more than 30 articles in leading national and international journals. He is also the author of the book *Hospital Service Quality* and is a frequent writer for trade magazines like *Pharma Biz*, in which he expresses his views on the current trends and business practices in the pharmaceutical and healthcare industry. He is Dean, School of Pharmaceutical Management at IIHMR University, Jaipur. His core research areas are behavioural intervention strategies using data analytics and IoT. At IIHMR, he is also actively engaged in various capacity-building programmes designed for the pharmaceutical industry for functional areas like Sales and Marketing.

Sunita Nigam, PhD is an assistant professor at the School of Health and Hospital Management, IIHMR University, Jaipur, India. She has a background in Biometrics and Nutrition with a Computer Science Diploma. She has more than two decades of experience in the health sector. Being involved in all kinds of large- and small-scale surveys, she has gained expertise in health survey research and programme evaluations. She has coordinated several research and evaluation studies independently, in different parts of the country. She has contributed in evaluations, police studies and was one of the first to do a survey on street children. Her areas of interest are programme evaluation, gender, women empowerment, reproductive health, STDs and HIV/AIDS, NCD diseases, nutrition, and NGO development. She has significant experience in scientifically conducting Qualitative and Quantitative research.

Sandeep Patyal is working as a project fellow in the Department of Commerce, University of Jammu. He has presented a number of papers in various national and international conferences and has also authored papers in various reputed journals.

Vijay Pereira is an associate professor of Strategic and International HRM at Khalifa University, UAE. Pereira was the Associate Dean (Research) at the Australian University of Wollongong, Dubai campus, prior to joining KU. He is Associate Editor (Strategic Management and Organization Behavior) of *Journal of Business Research*. Pereira is also a visiting scholar at Manchester and Portsmouth Universities, UK. He has expertise in consulting, industry, and academia, globally. Pereira also has ten books to his credit. He is currently on the editorial and advisory board for the journals *Production and Operations Management*, *International Journal of HRM*, *Asia Pacific Journal of Management*, *Asian Business & Management Journal* and *South Asian History and Culture*.

Sunita Raina is an assistant professor and Head of the Department of Commerce in Government Degree College Kathua. Earlier, she was a lecturer in Commerce and Management Sciences, Institute of Management Sciences, Jammu after receiving her doctorate in service marketing from the University of Jammu in 2008.

During her academic career, she has published two research papers. She has also written four book chapters in edited volumes, and presented six papers and attended several workshops. She regularly participates in various state level seminars, conferences and workshops. She is a member of Board of Studies of Commerce.

Asha Rani is a PhD scholar in the Department of Commerce, University of Jammu. Her areas of interest are Strategic Human Resource Management and Organisational Behaviour. She has written papers in international journals of repute.

Ritika Sambyal has teaching and research experience of seven years. She earned her M.Phil (Commerce) from University of Jammu and is pursuing her doctoral degree from the same university on "Examining the Outcomes of Corporate Social Responsibility (CSR) Initiatives of Companies in Punjab State". Her research is acknowledged in national and international journals. She has presented research papers in various national and international conferences.

Saguna Sethi is a PhD scholar working on "Effectiveness of Internal Marketing Mix Initiatives in the Higher Education Sector: A Study of Employee Attitude" in the Department of Commerce, University of Jammu. She has received her M.Phil on "Examining the antecedents of employee silence" from the same university in 2016. She has participated in various national workshops and also presented papers in national as well as international seminars and conferences.

Mamta Sharma is working as a lecturer in Udhampur Campus, University of Jammu (Jammu & Kashmir), India. Her area of interest is Financial Management. She has authored two books and seven papers in reputed national and international journals. She has also attended international and national conferences.

Poonam Sharma, PhD is a school teacher at the Jammu and Kashmir State Board of School Education, Jammu, India. Her area of specialisation is mentoring. She has written four papers in international journals. One of her papers is published in the book *Business Dynamics: Contemporary Issues and Challenges.*

Balan Sundarakani, PhD is an associate professor and program leader for the Master of Science in Logistics program in the Faculty of Business at the University of Wollongong in Dubai (UOWD). He has 12 years of teaching and research experience in the area of Supply Chain Management in various universities including the National Institute of Technology (NIT), Tiruchirappalli, Indian Institute of Technology (IIT), Roorkee, Hindustan University, Chennai and the National University of Singapore (NUS), Singapore and the Chair of Logistics Management in Swiss Federal Institute of Technology (ETH), Zurich, Switzerland. Sundarakani has authored 102 research papers which were published as book chapters and in refereed journals of international standing.

Kodati Viyyanna Rao is the former Vice-Chancellor of Acharya Nagarjuna University and In-charge of Krishna & Vikrama Simhapuri Universities, Andhra Pradesh. Rao is Acting Director of South East Asian Group of Institutions, Bangalore. He has 40 years of teaching experience in the field of Commerce and Management Studies and has attempted many innovations in Teaching and Research.

His research interests are focused on Public Enterprises, Advanced Banking and Finance and Academic Administration. A prolific writer, he has published 170 research papers in national and international journals. He has authored 20 books in the areas of Financial Management, Industrial Economy and Banking. The book entitled *Management of Schools in India—A Functional Analysis* which he co-authored with Dr. N. P. Prabhakara Rao and the book entitled *Socio-Economic Conditions of Handloom Workers*, with Dr. G. Naga Raju, were published by Lombart Academic Publishing, Germany.

He received Best Research Paper Award twice from the Institute of Company Secretaries of India, New Delhi, in 2002 and 2005 and Best Teacher Award from Government of Andhra Pradesh in 2009 and Outstanding Excellence in Academic Leadership Award from Nexus Eclat Honours Society, Maryland, USA in 2012. He received Life Time Best Research Award from Acharya Nagarjuna University and several Best Research Paper awards.

List of Figures

List of Tables

CHAPTER 1

Sustainable Business Practices for Rural Development

Intellectual Capital Perspective

Hardeep Chahal, Vijay Pereira, and Jeevan Jyoti

1 Introduction

Eradicating poverty, protecting the planet and ensuring prosperity for all are the United Nation's new Sustainable Development Goals (SDGs). In 2015, all UN members pledged their commitment to the 2030 Agenda for Sustainable Development. All member countries are expected to frame their respective development agendas on 17 goals relating to poverty, health, education, women, water, energy, economy, infrastructure, inequality, habitation, consumption, climate, marine systems, ecosystems, innovation, sustainable cities, institutions and overall sustainability that is to strengthen the means of implementation and revitalise the global partnership for sustainable development (UN n.d.).

H. Chahal • J. Jyoti (✉)
Department of Commerce, University of Jammu, Jammu,
Jammu and Kashmir, India

V. Pereira
Department of Humanities and Social Sciences, Khalifa University,
Abu Dhabi, United Arab Emirates
e-mail: vijay.pereira@port.ac.uk

H. Chahal et al. (eds.), *Sustainable Business Practices for Rural Development*, https://doi.org/10.1007/978-981-13-9298-6_1

Although the world is rapidly urbanising yet most of the world's poorest population lives in rural areas. Sound government policies are required to achieve SDGs in the context of rural areas in particular and urban areas in general. However, for business enterprises, sustainable development also refers to adopting business strategies and activities that will meet the needs of the enterprise and its stakeholders while protecting, sustaining and enhancing the human and natural resources that will be required for the future. All this, in turn, needs to be implemented through, or perhaps complemented by, programmes and projects that deliver goods and services to rural people through the public sector, private sector or non-governmental sector. Rural development initiatives require developing leaders and networking with those leaders to create rural economic vitality. The first step in creating this economic vitality is building an agreement that focuses efforts on significant activities based on community assets. In this context, Young (2010) surmised that rural businesses often embed locally, through personal networks, as a part of their survival strategies. The economic functions of rural areas have evolved considerably in recent decades. Among various economic activities, agriculture still plays a central role despite its declining importance worldwide economically and in the labour market. Agricultural intensification is common on "good land" or in more prosperous regions, whereas agricultural decline or abandonment is common in poorer or more marginal land.

Agriculture prosperity has been influenced by demographic and economic changes, which have reduced the dependency of rural inhabitants upon farming and other primary industries (Satterthwaite et al. 2010). New industrial and service activities have emerged that to a certain extent have started replacing agriculture activities (Baldock and Lowe 1996; Baldock et al. 2001). A gradual shift from a sectoral to a spatial focus is affecting the rural economy resulting in a progressive detachment from the exclusive production of food and fibre and in a concomitant increasing reliance on the service economy, tailored to the new requirements of the urban society (Baldock and Lowe 1996). Although some of the requirements (e.g., housing, business relocation and certain kinds of outdoor recreation) cannot be easily reconciled with the carrying out of farming activities, agriculture can still have a pivotal and catalysing part in meeting other equally relevant demands placed on the countryside such as rural tourism, the preservation of rural landscapes and traditions, environmental education and the production of healthy native food.

Many social scientists are finding it difficult to develop a rural development model that is capable of bringing a paradigm shift (Ploeg et al. 2000). Polidori and Romano (1996) remarked that sustainable rural development is a process of multidimensional change affecting rural systems. Economic growth, improvement of social conditions and conservation of natural values are important for sustainable rural development and should be induced according to a bottom-up approach through a participated and sustainable use of local endogenous resources (environment, labour force, knowledge, patterns of production, consumption and communication) (Pugliese 2001). Public-private participation has become a widely advocated methodological principle to work for the overall sustainable development of the rural regions (Leeuwis 2000).

Looking into the aforementioned rural development issues, this chapter highlights the role of intellectual capital, which can also be considered as the DNA for rural business growth in particular and rural development in general. The following section discusses, in brief, the concept of intellectual capital and its dimensions.

2 Intellectual Capital

Intellectual capital is recognised as one of the critical sources for an organisation to gain competitive advantage (Barney 1991). Its role in sustaining this competitive advantage is recognised by various researchers in service and manufacturing organisations in both developed and developing countries. Rapid changes in customer expectations and technology, unpredictable customer behaviour, hyper-competition are acknowledged as significant factors leading to the rise in the research on intellectual capital for sustaining business performance (Chahal and Bakshi 2015, 2016). The concept of intellectual capital has been drawing attention globally with its claim to be an effective competitive tool for value creation. However, there is no such single definition that is shared by the researchers. Edvinson (1997) and Sullivan (1999), the two major contributors in the development of intellectual capital, considered intellectual capital as a composite of human competencies, knowledge, ability, skills, experience and customer relationships that provided the competitive edge to an organisation. Specifically, Edvinson (1997) defined intellectual capital as a set of intangible assets that includes resources, competencies and capabilities that enhances organisational performance through an increase in value creation. Similarly, Sullivan (1999) considered intellectual capital in terms of competencies and relationships and defined it as an intangible asset resulting from (intangible) resources. This concept was also supported by Ulrich (1998). From a financial perspective, intellectual capital reflects the difference between the market value and the book value of the firm (Sveiby 1997). Further, Roos and Roos (1997) defined intellectual capital as a combination of significant capabilities and structures that generate organisational value. However, all these definitions have been derived from a narrower perspective. Bontis, in 2004, expanded the concept of intellectual capital from a broader perspective and defined it as "the hidden values of individuals, enterprises, institutions, communities and regions that are the current and potential sources for wealth creation. These hidden values are the roots for nourishment and the cultivation of future well-being". Hence, the role of intellectual capital in the sustainable development of rural sector can be considered from the role played by different organisations and stakeholders that include employees, suppliers, customers, society, government and so on.

Human capital, relational capital and structural capital, the three components of intellectual capital, are well accepted in the literature by majority of the researchers such as Chahal and Bakshi (2015), Shih et al. (2010), Cabrita and Bontis (2008), Kavida and Sivakoumar (2009), Bontis et al. (2000), Edvinson (2000), Sullivan (1999), Stewart (1997) and Roos and Roos (1997). Hsu and Wang (2010) have also remarked that human capital is a basic component of the intellectual capital process that drives the other two components of the intellectual capital—structural capital and relational capital. A brief description of these three components has been provided in the following sections.

Human Capital

Human Capital is defined as an organisational potential especially in terms of know-how, capabilities, skills and expertise of the employees and is most effective for sustaining competitive advantage because of its source of innovation and strategic renewal characteristics (Bontis 1998; Edvinson and Malone 1997). According to Halim (2010), human capital is what employees bring into the value-adding processes and encompasses professional competence, employee motivation and leadership ability. Edvinson and Malone (1997) and Isaac et al. (2010) remarked that human capital reflects the organisational knowledge generated from its present employees, which does not remain in the organisation when they (employees) leave the organisation and hence, it possesses transferability characteristics. Edvinson and Malone (1997) have related human capital to the capabilities and competencies of organisational employees that generate value. Specifically, they have equated human capital with the collection of attitudes, skills, competencies and abilities of people which contributes to organisational productivity. In simple words, it is viewed as how effectively an organisation uses its human resources' experience, learning, skills, education, competence and creativity to create organisational value (Bakshi 2014).

The factors affecting human capital value include competence, attitude, agility, imagination and integrity (M'Pherson and Pike 2001), know-how, problem solving, decision-making and learning (Edvinson 1997; Robinson and Kleiner 1996), employees' collective competence, capabilities and brainpower (Seetharaman et al. 2004), genetic inheritance, education, experience, attitudes about life and business (Hudson 1993). Among all the varied demographic factors, education plays an important role in enhancing competency and capability vis-a-vis human capital. The quality and nature of formal educational programmes in the initial stages along with enhancement in learning and experience also paved the way for the development of intellectual capital.

Relationship Capital

Relationship capital is also considered as external capital or customer capital. According to researchers such as Shih et al. (2010) and Cabrita and Bontis (2008), relational capital is the knowledge embedded in relationships with customers, suppliers, industry associations or other stakeholders that influences organisational life, creates value and adds to enhanced organisational functioning. Alternatively, it denotes the strength of organisations' relationship with external stakeholders. Welbourne (2008) considered relational capital in terms of development and maintenance of quality relationships with organisations, individuals or groups that influence or impact the value creation of businesses. However, owing to its external nature, knowledge embedded in relationship capital is very difficult to codify (Bontis 1998).

Structural Capital

Structural capital is also considered as internal capital by scholars such as Edvinson and Malone (1997), Stewart (1997) and Sveiby (1997). Structural capital relates to organisational structure and systems, which support employees' productivity and that remain in the company even when employees leave an organisation (Edvinson and Malone 1997; Roos et al. 1997). It is described as an enabling structure comprising systems, procedures and processes which enhance employees' capability to do work and thus facilitates the organisation in harnessing intellectual capital (Bontis 1998; Seetharaman et al. 2004). It includes hard assets such as building, machinery, infrastructure of the firm and soft assets such as policy framework, guidelines, rules, regulations, employees' relationships andso on. The knowledge is embedded within the routines of an organisation and deals with the mechanisms and structures of the organisation that supports employees to contribute effectively for optimum intellectual performance (Bontis 1998). Later, Bontis et al. (2000) described structural capital as non-human storehouse of knowledge, which includes databases, organisational charts, process manuals, strategies, routines and any other thing whose value to the organisation is higher than its material value. They also remarked that if an organisation has poor systems and procedures, the overall intellectual capital of organisation will not reach its fullest potential.

3 Role of Intellectual Capital in Rural Development

This book is about the role of intellectual capital in rural development that has been examined from human, relationship and structural capital perspectives. The contribution of intellectual capital towards rural development in the organisations is assessed from two perspectives: first, the role of the employees working at the top, middle and low levels of management and second is the rural employees' contribution to the organisation's performance. The contents of each section described briefly are detailed further.

Human Capital

This part of the book emphasises the employees' behaviour and attitudes in the conduct of organisational functioning. Specifically, it encompasses studies that highlight the role of organisation citizenship behaviour and employees, express ability and self-efficacy in enhancing employee participation and organisational productivity.

Relationship Capital

This section focuses on the organisational relationship with external customers that include customers, suppliers, distributors and so on. The effectiveness of rural services in the context of services such as healthcare, education, financial and so on have been evaluated to design strategies for enhancing satisfaction.

Structural Capital

This part highlights the evaluation of soft and hard measures and their impact on the financial and non-financial parameters of the organisations. Specifically, government schemes such as Mahatama Gandhi National Rural Employment Guarantee Act (MGNREGA), Self Help Group (SHG), Foreign Direct Investment (FDI) and so on have been critically examined to evaluate their impact on rural development.

4 Conclusion

This chapter discusses brief, the role of intellectual capital in general and in context of rural development in particular. Intellectual capital is considered as a significant factor in the socio-economic development of the country and the region. Its three components namely human capital, structural capital and relational capital play a pivotal role in enhancing firm performance and contribute to the rural economy.

References

Bakshi, P. (2014). Effect of Intellectual Capital on Competitive Advantage and Business Performance: Role of Innovation and Learning Culture in Banking Sector. PhD thesis, University of Jammu, Jammu.

Baldock, D., Dwyer, J., Petersen, J., Lowe, P., & Ward, N. (2001). *The Nature of Rural Development: Towards a Sustainable Integrated Rural Policy in Europe.* Report to WWF European Policy Office and the Land Use Policy Group, IEEP, London (http://www.lupg.org.uk/pubs/) ISBN 1-873906-40-4.

Baldock, D., & Lowe, P. (1996). The Development of European Agri-environment Policy. In M. Whitby (Ed.), *The European Environment and CAP Reforms* (pp. 8–25). Wallingford: CAB International.

Barney, J. (1991). Firms, Resources and Sustained Competitive Advantage. *Journal of Management, 17*(1), 99–102.

Bontis, N. (1998). Intellectual Capital: An Exploratory Study that Develops Measures and Models. *Management Decision, 36*(2), 63–76.

Bontis, N. (2004). National Intellectual Capital Index. *Journal of Intellectual Capital, 5*(1), 13–39. https://doi.org/10.1108/14691930410512905.

Bontis, N., Keow, W. C., & Richardson, S. (2000). Intellectual Capital and Business Performance in Malaysian Industries. *Journal of Intellectual Capital, 1*(1), 85–100.

Cabrita, M., & Bontis, N. (2008). Intellectual Capital and Business Performance in the Portuguese Banking Industry. *International Journal of Technology Management, 43*(1), 1–26.

Chahal, H., & Bakshi, P. (2015). Examining Intellectual Capital and Competitive Advantage Relationship Role of Innovation and Organizational Learning. *International Journal of Bank Marketing, 33*(3), 376–399.

Chahal, H., & Bakshi, P. (2016). Development of Intellectual Capital Scale for Indian Banks. *Vikalpa: The Journal for Decision Makers, 41*(1), 61–73.

Edvinson, L. (1997). Developing Intellectual Capital at Skandia. *Long Range Planning, 30*(3), 266–373.

Edvinson, L. (2000). Some Perspectives on Intangibles and Intellectual Capital. *Journal of Intellectual Capital, 1*(1), 12–16.

Edvinson, L., & Malone, M. (1997). *Intellectual Capital: Realising Your Company's True Value by Finding its Hidden Roots.* New York: Harper Business.

Halim, S. (2010). Statistical Analysis on the Intellectual Capital Statement. *Journal of Intellectual Capital, 11*(1), 61–73.

Hsu, L.-C., & Wang, C.-H. (2010). Clarifying the Effect of Intellectual Capital on Performance: The Mediating Role of Dynamic Capability. *British Journal of Management, 23*(2), 179–205. https://doi.org/10.1111/j.1467-8551.2010.00718.x.

Hudson, W. J. (1993). *Intellectual Capital: How to Build It, Enhance It, Use It.* New York: John Wiley & Sons Inc.

Isaac, R., Herremans, I., & Kline, T. (2010). Intellectual Capital Management Enablers: A Structural Equation Modelling Analysis. *Journal of Business Ethics, 93*, 373–391.

Kavida, V., & Sivakoumar, N. (2009). Intellectual Capital: A Strategic Management Perspective. *The IUP Journal of Knowledge Management, 7*(6), 55–69.

Leeuwis, C. (2000). Reconceptulization for Sustainable Rural Development: Towards a Negotiation Approach. *Development and Change, 30*, 931–959.

M'Pherson, P. K., & Pike, S. (2001). Accounting, Empirical Measurement and Intellectual Capital. *Journal of Intellectual Capital, 2*(3), 246–260.

Ploeg, J. D., et al. (2000). From Practices and Policies Towards Theory. *Sociologia Ruralis, 40*(1), 391–408.

Polidori, R., & Romano, D. (1996). Agricoltura e sviluppo, dattiloscritto, lavori preparatori di 'Agricola 96' Conferenza Provinciale dell'Agricoltura, Firenze 18 novembre.

Pugliese, P. (2001). Organic Farming and Sustainable Rural Development: A Multifaceted and Promising Convergence. *Sociologia Ruralis, 41*(1), 112–129.

Robinson, G., & Kleiner, B. (1996). How to Measure an Organization's Intellectual Capital. *Managerial Auditing Journal, 11*(8), 36–39.

Roos, G., & Roos, J. (1997). Measuring Your Company's Intellectual Performance. *International Journal of Strategic Management, 30*(3), 413–426.

Roos, J., Roos, G., Dragonetti, N., & Edvinson, L. (1997). *Intellectual Capital: Navigating the New Business Landscape*. London: Macmillan Press.

Satterthwaite, D., McGranahan, G., & Tacoli, C. (2010). Urbanization and Its Implications for Food and Farming. *Philos Trans R Soc Lond B Biol Sci., 365*(1554), 2809–2820. https://doi.org/10.1098/rstb.2010.0136.

Seetharaman, A., Lock, T., Low, K., & Saravanan, A. S. (2004). Comparative Justification on Intellectual Capital. *Journal of Intellectual Capital, 5*(4), 522–539.

Shih, K., Chang, C., & Lin, B. (2010). Assessing Knowledge Creation and Intellectual Capital in Banking Industry. *Journal of Intellectual Capital, 11*(1), 74–89.

Stewart, T. (1997). Intellectual Capital: The New Wealth of Organisation. *Business Digest, 300*, 1–10.

Sullivan, P. H. (1999). Profiting from Intellectual Capital. *Journal of Knowledge Management, 3*(2), 132–143.

Sveiby, K. (1997). *The New Organisational Wealth: Managing and Measuring Knowledge-based Assets*. New York: Berrett-Koehler.

Ulrich, D. (1998). Intellectual Capital Equals Competence X Commitment. *Sloan Management Review, 39*(2), 15–27.

United Nations. (n.d.). Sustainable Development. Retrieved from http://www.un.org/sustainabledevelopment/sustainable-development-goals/

Welbourne, T. (2008). Relational Capital: Strategic Advantage for Small and Medium-sized Enterprises (SMEs) through Negotiation and Collaboration. *Journal of Business and Economics, 18*(5), 438–492.

Young, N. (2010). Business Networks, Collaboration and Embeddedness in Local and Extra-local Spaces: The Case of Port Hardy, Canada. *Sociologia Ruralis, 50*(4), 392–408.

CHAPTER 2

Rural Development Through Sustainable Business Practices: Juxtaposition of Private and Public Initiatives

Kodati Viyyanna Rao

1 The Context

'India lives in villages' was the philosophy of Mahatma Gandhi—the Father of our nation. His concept of village economy was one where it was a self-sustaining, self-sufficient and self-propelling mechanism. He believed that the only way to develop India is to make 'village' the central instrument for all our development planning exercises. He demonstrated to the world through his Sevagram Centre, which is located near Wardha in Maharashtra, as to how a village reconstruction can take place. His 18-point constructive programme is indicative of his philosophy of rural transformation which is based on broader themes such as swaraj, swadeshi, trusteeship, self-sufficiency, decentralization, equality and dignity of life (Bader 1995; Pyarelal 1977). His idea of village swaraj was based on complete independence, implying production of own food, own cotton and own cattle and all other necessities. He also entertained, during those years itself, the idea of 'exportable surplus' by suggesting the farms to

K. Viyyanna Rao (✉)
SEA Group of Institutions, Bangalore, India

H. Chahal et al. (eds.), *Sustainable Business Practices for Rural Development*, https://doi.org/10.1007/978-981-13-9298-6_2

grow many crops, if more land is available. It appears that after having operated the economy for over seven decades, we are still stuck with a 'development model' that enables us to forge ahead in the international arena. It appears that we are keeping faith in the tenets of Mahatma Gandhi in this regard (Gandhi 1959).

2 Sustainable Business Initiatives for Rural Development

In the context of the discussion on Sustainable Development, it would be interesting now to turn our attention to the business initiatives that help transform rural areas of our country. As a matter of fact, these practices pertain to a wide variety of areas such as:

- Eco-efficiency, mainly achieved by the delivery of competitively priced goods and services.
- Increased resource productivity to get more and more goods and services from lesser input materials or energy sources.
- Recycling and waste reduction and management, implying 'wealth from waste'.
- Transition to renewable sources of energy.
- Green technologies to improve the natural light and air and also reduce dependence on technologies that contribute to 'global warming'.
- Accountability for the environment.

All these initiatives are intended to protect or combat the areas of concern for sustainable development such as ozone depletion, reducing biodiversity, global warming, climate changes, eroding soil fertility, melting of glaciers, increasing sea levels and indiscriminate usage of toxic substances (UN Development Report 2015). Any of the business practices that express concern over these issues can be termed 'sustainable' and hence deserve a hearty welcome (Bentham 2009; Mill 1863). Against this setting, an attempt has been made in the following pages to examine two business practices that help achieve these objectives and contribute immensely towards rural transformation. Typically one is in the domain of private sector and had been championed by ITC Ltd in the name of 'e-Choupal' which was launched in June 2000. The other one is being introduced and uses the technology by the name of 'e-NAM' in the public sector domain.

3 The Case of e-Choupal of ITC

For the first time in India, ITC (a business conglomerate in India known for its tobacco products and now a multi-product company including hotels as its assets) has conceived the idea of creating an electronic platform which assists farmers in marketing their produce and includes aquaculture too. The marketing of agricultural products has been dominated by the intermediaries and middlemen (Bhatia 2007). Farmers used to be at the mercy of these middlemen earlier.

The Process

The e-Choupal is an electronic facility within which ITC has established internet kiosks at selected villages. Each Choupal is run by a sanchalak who is a trained farmer. The computer system is placed in the house of the sanchalak who is provided with internet connection via phone or VSAT. Each facility is designed to serve an average of 600 farmers in the surrounding ten villages within a radius of five km. The farmers in these villages are advised to visit the Choupal for any kind of service, such as prices of agricultural products at different mandis, prices and availability of inputs, fertilizers and so on. Through this facility, farmers can sell their products directly to ITC as well. There will be a service fee for each transaction and the same would go to the sanchalak for taking care of the facility. The experiment which began in 2000 on a limited scale has now grown mighty with about 6100 e-Choupals covering 40,000 villages and 4 million farmers in ten states of the country.

The Value Chain Model

The creators of e-Choupal, ITC, believe that the e-Choupal would help in creating a new value chain in the agriculture sector which was exploited by money lenders and middlemen. One can imagine that this model would enhance competitiveness of Indian agriculture and trigger a virtuous cycle of higher productivity, higher incomes to farmers thus unleashing the latent demand for industrial goods necessary for triggering the growth trajectory of Indian economy (Bowonder et al. 2002). The system is designed to be a direct marketing channel as it is linked to the mandis for price discovery. It also helps in eliminating wasteful intermediation and multi-level handling. The ITC's maiden attempt has turned out to be a trendsetter in linking technology to rural development through agriculture. It could contribute to the growth of local leadership in villages, sharing of ownership of assets, opportunity to know many facets of the agriculture—not only prices, but the diverse practices adopted in agricultural operations—and finally leading to the enhancement in income levels through productivity.

4 From Concept to Completion

It is said that the senior executives had gathered together to create and leverage an electronic market place as a 'business model' to enhance rural marketing opportunities to the farmers and also help the agri-business division of the ITC. It is known that ITC is a major exporter of soyabean. It used to buy the quantity mainly from the local producers. As is the case with many rural markets, the produce is of poor quality and poor grading with a high cost of intermediation. When the company wanted to compete on an international scale, these obstacles needed to be overcome. They realized that there is a need to create a suitable physical, social and institutional infrastructure coupled with the acumen of already existing intermediaries. Very appropriately, ITC did not intend to eliminate the intermediaries but at the same time designed the scheme to leverage their strengths for the success of their model. Besides, the company also attempted to introduce 'service element' in its activities and improve the customer centricity, by focusing on the introduction of new lots, new varieties, new packaging and non-standardized orders.

As can be seen from the model, the major strength of the model lies in the creation of 'network', that is, the ability to connect to all kinds of players in the markets, whether big or small, thereby eliminating the need for hierarchal middlemen through a low-cost transaction mechanism. The model helped in transforming the existing market micro-structure in the following ways:

- Enhancing farm productivity through (a) dissemination of latest information on climate changes, weather forecasts and other factors, (b) modern methods of farming and (c) through the supply of quality inputs such as seeds, pesticides, fertilizers and so on.
- Improving upon the price discovery mechanism through constant updating of price information available at the local, national and international markets. Also helping in the trends pertaining to supply and demand for agro-products at the local, national and international

markets and sharing of expert opinion on the current and future trends with respect to demand, supply and prices.
- Minimizing transaction costs by purchasing the produce at the farmer's field/house and by following all the transparent practices with respect to checking the quality, grading, price and weight.

Thus, the experiment of ITC in the agro-marketing/rural marketing remained a novel and dynamic initiative in terms of (a) creating local leadership (in the form of sanchalaks), (b) sharing of agro-information in the best possible manner, (c) negating the role of middlemen (especially breaking the hierarchical and tyrannical system and market micro-structure), (d) introducing transparency in every deal, and finally (e) showing to the entire world the best possible way of integrating Information and Communication Technology for the upliftment of rural lives.

5 The Case of e-NAM

e-NAM stands for electronic National Agricultural Market that acts as an electronic trading platform. This is intended to bring into the network all the existing Agricultural Produce Marketing Committee (APMCs) spread across the country. As per the information available, around 7000 APMCs are operating in the country. The major objectives of e-NAM are as follows:

- To create a unified national market for all agricultural commodities.
- To act as a single-window service for all the APMCs across the country for any agricultural information.
- To eliminate hindrances and inconsistencies in the operation of APMCs.
- To help in the free flow of information on production, prices, markets and enhance transparency in the agricultural system.
- To ensure free flow of agro-produce across states and seek to eliminate inter-state barriers to trade in agro-products.
- To help in the streamlining of procedures followed by different states.
- To assist in the process of real-time price discovery based on actual demand and supply.

The key stakeholders in the system include farmers, traders, buyers, processors and exporters. The e-NAM is expected to provide benefits to all these stakeholders in terms of competitive prices, reduced transaction costs and better grading and quality. There is a long list of 90 varieties of agro-produce varying from cereals to fruits, vegetables, seeds, pellets, pulses, spices and bamboo.

Operational Guidelines

The Operational Guidelines issued for the stakeholders included:

- Scheme's design.
- Eligibility criteria for availing assistance under the scheme.
- Nature of assistance being provided.
- Process followed for release of the assistance.
- Role of various agencies associated with the system.

As per e-NAM's guidelines, it started operating from 1 July 2015 with a budget allocation of Rs 200 crores. Small Farmers' Agro-business Consortium (SFAC) was identified as the implementing agency. Though e-NAM is a virtual trading platform, it has also to care for the physical flow of the commodities at the APMCs at the ground level. SFAC is expected to coordinate all these issues. In order to be able to avail the assistance under the scheme each state/UT is supposed to put the following in place:

1. Single trading licence to be valid across the State.
2. Single-point levy of market fee across the State.
3. Creation of necessary infrastructure for e-auction/e-trading as a mode of price discovery.

Though SFAC was identified as a lead agency, all the state and UT level institutions are directed to involve themselves in the operation. M/s Nagarjuna Fertilizers and Chemicals Ltd (SFCL) was appointed as the strategic partner (SP) for a period of five years with effect from 23 December 2015. The SP should design, develop, test, implement and maintain the platform to the satisfaction of the stakeholders.

Evaluation of the Model

It is evident from the previous discussion that the scheme had been introduced with fanfare. By now, it has been in operation for about two and a half years. The coverage of the scheme has been poor and it could cover only about 455 mandis out of 7000 by November 2017 and could gather only 45.4 lakh farmers out of about 1300 lakh. As indicated earlier, for the scheme to be successful, every state and UT needed to partner with the centre and suitably amend their laws to pave way for the unified (single) licence and unified (single) fee. So far only 13 states have done the necessary. It is also being reported that in no state, the platform is fully functional for want of necessary equipment such as sorting/grading machines, testing machines, besides the shortage of computer systems and connectivity.

Unfortunately, in some of the states, mandis were closed as an instant reaction to the new mechanism. Both the traders and the commission agents were conspicuous by their absence in carrying out their operations. The reasons cited by them included the following:

- First, the traders generally give an advance to the farmers for buying inputs and carrying out the agricultural operations, mostly with the condition that the farmer would sell the produce only to him. Under the new model, if the farmer sells to some other trader through e-NAM, the recovery of his loan is said to be an issue. Usually, there used to be a heavy bargain and distress selling or even a tie-up sale earlier. Now all these have been withdrawn.
- Second, under the new model, the sale amount should be credited to the bank account of the farmer on the same day by the trader. In the past, the trader used to pay at his convenience after 15 days to one month, mostly after he sold the produce and received the money for it. He used to invest less or nil, but was involved in trading. Now, every trader has to invest money to participate in the buying process.
- Third, under the new model, the quality of the produce is to be decided by the labs established for this specific purpose and these labs are attached to the mandi, depending upon the type of product. The lab concerned would issue the certificate based on the quality. There shall not be any contention on this. It is hoped that this would help both the farmer and the trader. Since the farmer is not so educated and does not have the relevant testing equipment, he would have the advantage of a 'Quality Certification'. For the trader being away from the market and not able to physically verify/check the produce, it can go by the quality certificate issued by the testing lab and can buy the commodity from anywhere in the country. However,

there are two issues: one, the testing labs were not established in majority of the mandis. Two, traders are averse to this kind of system as they had the advantage of fixing quality on their own much to the dismay of the farmer.

- Fourth, there is an apprehension among the farmers and the traders equally that their identities are now known nationwide, since they are required to submit all the details such as aadhar card number, bank accounts, volume of trade and many more. Earlier, all this information remained confidential, thus having the scope to park unaccounted money resulting from various transactions. The commission received by the agents too attracts GST as per the new rules.
- Fifth, at several mandis, there prevailed an unofficial and unauthorized collection of toll/fee/charge by the agents as well as the executive body of the markets. Under the new system this will not be possible.

For these and many other reasons, the scheme proved to be a 'damp squib' (Das 2017; Rajalakshmi 2017). The very interesting thing is that in the state of Andhra Pradesh, all the prominent market platforms got closed as if there was a holiday declared. Andhra Pradesh is one of the agro-based states known for its production of commercial crops like chillies, turmeric, cotton, tobacco and certain kinds of spices such as ginger, pepper, mustard and coriander. Andhra Pradesh has been the major producer of these crops and the purchase of these items takes place at a national scale. The state implemented e-NAM in 22 market yards. Immediately after the implementation, major platforms located at Guntur (for chillies) and Duggirala (for turmeric) got closed because of a lack of trade. For instance, in the Guntur market yard, every month (during the season) on average, about five lakh bags of chillies used to be traded. This area is also known for the presence of the highest number of 'cold storages'. After e-NAM, the trading got limited to just 617 bags, meaning that there was no trading at all. These examples portray the inbreeding prevailing at the markets/mandis and the unholy nexus among buyers, commission agents and sometimes officials. This chain needs to be broken down. The best effective solution is to press the government departments and agencies into service and start buying. In those places where undertakings like Marketing Federation (MARKFED), Food Corporation of India (FCI), Cotton Corporation of India (CCI) and so on are jumping into action, the problem is getting resolved. But one difficulty with these organizations is that they can buy goods only at the support price declared by the Government of India. We are aware that the support price is not remunerative price to the farmer. Therefore, there is an imminent need to devise a mechanism to toe all the buyers and agents on the one hand and also to activate the government organizations and departments to plunge into action, wherever warranted.

It is a matter of gratitude that the Government of India is committed to strengthen the infrastructure. It was announced on 25 December 2017 that the government intends to enhance the infrastructure grant from Rs 30 to 50 crores to each mandi and see that all the mandis are electronically connected. It is also appreciable that the government is proposing to increase the list of committees to more than a hundred.

6 Lessons from This Experience and the Challenges Ahead

An attempt has been made in this section to draw lessons from the two case studies and unveil the challenges ahead in creating suitable and sustainable models in the Indian context.

- It is evident from the ongoing discussion that it is not very easy to design and devise creative models for achieving sustainable development. Increasing temperatures of the earth are a mammoth challenge for the people and policymakers. In India, the effort is very fragile and weak because sustainable development implies persistent action about the saving of the environment. Almost everything is possible in India, like running a 30-year-old vehicle on the highways, horns without a decibel limit, drawing of water from the ground level and depleting it and pollution of all kinds such as air, water, sound and ground through Open Defecation Free (ODF). It would be better to talk less about the transfer of technology and the inflow of FDI into India. All the technologies that have become redundant at their home country are finding a way into India, especially of chemicals, paints and pharma. We do not even leave wells, tanks, lakes and rivers alone. Yet, thousands of crores are earmarked for cleaning them under special projects like the cleaning of Ganges with a budget of about Rs 20,000 crores. Real estate is permitted at all places including tanks, lakes and river courses. Mining is indiscriminate including 'sand' in the rivers. For our government, prevention appears to be uneconomical and less profitable than restoration. Therefore, one is required to be innovative with these odds.

- Turning our attention to agriculture, since both the case studies pertain to this major sector, our position is still heavily subjected to the vagaries of nature. It holds the future of about 119 million farmers and about 144 million agricultural labourers. When the monsoon is fruitful, we witness bumper crops (as evidenced in 2016–2017 with an estimated cereal food grain production of 273 million tonnes), and vice-versa in other seasons. One interesting observation is that India with its vast tract of arable land faces drought at one or the other areas. For instance, at the best possible production year of 2016–2017, around 30 per cent of the districts needed to be declared drought-prone, and needed to be provided with a bail-out package in terms of loan/interest waiver, subsidized inputs and so on. Additionally, there is still an excessive dependence on agriculture by the working class. As per the recent reckoning, about 60 per cent of the Indian population is still dependent on agriculture for their livelihood; yet its share in the GDP is at 14–15 per cent. India has the second largest arable land (second only to USA) with about 159.7 million hectares. Of which only 51.72 per cent is the gross irrigated crop area and the net irrigated is about 60 per cent of the same only. Interestingly, of the above 160 million hectares of cultivable land, only about 35 per cent of the same can be reliably irrigated. The rest is dependent on groundwater (about 39 million hectares) and the rest (about 65 million hectares) is based on the village tanks, culverts and the like. Thus, we need to be creative in these areas to be sustainable in this most widespread sector of occupation (Rajvanshi 2002).
- One should have known by now that the burning of the forests for purposes such as cultivation, wood, real estate and others. is damaging to the environment. But states like Punjab developed a dangerous tendency which was the burning of fields. Despite the ban on crop burning, farmers in the state of Punjab continue to set fire to the paddy residue to make way for the next crop. To contain this practice, the National Green Tribunal (NGT) in 2015 had issued an order to stop the burning of stubble, since thick smog enveloped the northern skies, adding to the woes of the Delhi climate that is already choked due to vehicle and industrial air pollution. The NGT has also imposed fines for indulging in such practices, varying from Rs 2500 to farmers owning less than two acres of land and Rs 5000 to those

who own above two but below five acres and Rs 15,000 to those owning above five acres. The NGT instructed the Government of Punjab to take stern action against those involved in such practices. However, nothing could stop the farmers from this environmentally hazardous practice. What is more, a group of farmers belonging to an organization called Bharatiya Kisan Ekta is engaged in guarding the fields that are set on fire to not allow the government officials from reaching the area. In a more recent order dated 13 May 2019, the National Green Tribunal has called for a Status Report on the Action taken by the Government of India curbing stubble burning in different states.

These incidents make us think seriously about sustainable practices like the generation of biomass energy using the stubble. Going by the data of the Government of Punjab, paddy is said to be grown in Punjab in about 30 lakh hectares, resulting in about 19.7 millions of paddy straw every season, which is to be cleared from the fields to make way for the next crop, that is, wheat. It is estimated that about 70–75 per cent of this straw is burnt in the open fields itself and considered as the cheapest and quickest way of getting rid of the residue. As per the available information, only about 4.3 million tonnes of this straw is at present consumed in the industry for manufacturing paper, cardboard, animal fodder and so on. Companies like ITC that also deal in cardboard production can come out with the newer technologies and processes. In addition, bagasse generated from sugar factories is also an issue. Like the farmers of Punjab, they too are involved in using the bagasse for heating; but they are helpless in controlling the emission of smoke or smog (*The Hindu* 2017).

- There is also the case of reconstruction of villages based on the model suggested by Mahatma Gandhi. In that model, we need to make every village self-sufficient, if not independent. At least as a contingent measure, the idea of late A.P.J. Abdul Kalam needs to be tried. He advocated for many years, even before he became the President of India, to focus on village reconstruction and the result was the scheme called Providing Urban Amenities to Rural Areas (PURA). The idea of his scheme was that it would bring together the experiences of both public and private sectors in the development of our rural areas. He also wished for the private sector to bring in investment besides experience. The public-private partnership (PPP) is the mode through which he wished to achieve some results in this area. His focus was on the following areas, connecting rural and urban living with the aim of rooting out the existing divide between them.

- Physical connectivity
- Electronic connectivity
- Knowledge connectivity
- Economic connectivity

It was expected that the scheme would operate as a self-sustainable and a viable model of service delivery in the rural areas with the core funding coming from the Central Government and the operational investment being provided by the private sector on considerations of economic viability. Dr Kalam was very elaborate, eloquent and emphatic of this idea that he also elaborated through his book titled *Target 3-Billion* co-authored by Srijan Pal Singh. Though the Central Government took up the matter with the Ministry of Rural Development (MoRD) and initiated a few pilot projects in select areas such as Basmath (Maharashtra), Bharthana (Uttar Pradesh), Gohpur (Assam), Kujanga (Odisha), Motipur (Bihar), Rayadurg (Andhra Pradesh), Shahpura (Rajasthan). Despite this, the scheme largely remained a non-starter.

In this context, the very idea of village reconstruction can be tried with the help of corporates. After the incorporation of the specific clause in terms of spending by corporates under the Corporate Social Responsibility (CSR), 2 per cent of their average profits of the preceding three years, may be entrusted to one village of their choice and under this responsibility, they may introduce schemes of business and non-business nature (Raju and Rao 2017). As per the 2011 Census, we have 640,867 villages and about 236,004 (36.8 per cent) of them are so small that the average size of the population is below 500. In contrast, we will have 1,704,319 companies registered in India by 31 October 2017. Even when we ignore and delete from the list of the non-active or non-functioning companies (said to be around 573,535) we will have 1,130,784 active companies; almost double the number of villages. It not as though it is mandatory and that each of the companies will be persuaded to be benevolent in this respect but even if 20–30 per cent of them respond and involve themselves like the ITC, the face of rural India would undergo a massive change. This should be envisioned on a 'revolutionary mode' (Goyal 2010).

After the stipulation of mandatory spending by the corporates under the CSR, companies in total have spent about Rs 8300 crores in 2015–2016 alone. This spending by companies/conglomerates just individually is so significant that there are companies like Reliance Industries (620 crores), TCS (446 crores), HDFC Bank (304 crores), Infosys (287 crores) and ITC (275 crores) on the top of the list, as per the latest information (Raju and Rao 2017). Going by the area of focus of the companies in discharging their responsibilities under this programme, one has to admit that some of the companies are highly innovative and apt to the concept of rural development. Following is the list of such interesting initiatives:

- *Water Security Initiative of Reliance Industries Limited (RIL)*: Wherein RIL has developed water harvesting structures for the communities living in dry and rain-fed areas and harvested about 447 lakh cubic metres of rain water.
- *Social forestry programme of ITC*: This project is meant to cover about 70,000 hectares of land covering 3720 villages, impacting 70,000 households.
- *Biodiversity project of Wipro*: This is intended to reduce carbon intensity and improve water retention. It contained four themes to cover (a) Butterfly Park, (b) Natural Water Body Designs, (c) Medical Circuit, (d) Wetland Park.
- *Ankur of HUL*: It is a centre started by the company to provide special education to the differently abled children in Assam.
- *Making the World Greener initiative of Airtel*: This project is meant to reduce the carbon footprint in the selected areas such as Nigeria, Ghana, Zambia and Uganda. The company is operating this project in these African countries.

It is known that the 72nd and 73rd Amendments to the Constitution of India paved the way for the reform in local self-governance and made the provision for the transfer of funds to the Local Bodies (LBs) directly from the Central Government. Through these amendments, LBs were also entrusted and empowered with functions that were carried out independently in the form of 29 items by the panchayats and 16 items by the Nagarpalikas. This was dubbed as the 'Democratic Decentralisation' in the history of independent India.

The spirit of these amendments needs to be protected, both in terms of legislative and financial terms. It is true that the successive Finance

Commissions (FCs) are doing their bit in support of the strengthening of the LBs in rural India. There have been cases of aberration that have been reported in a few states which mentioned that the elections were not conducted properly. LBs are not allowed to spend money and withhold the grants received from the Centre under any pretext. This is said to be more pronounced in the case of LBs under the rule of opposition parties. If villages are to develop on a sustainable basis there is no other option except to implement the constitutional amendment in letter and spirit.

One more idea which appears to be very viable and everlasting is the formation of the Association of Non-Resident Villagers (ANRV), something like an alumni association of educational institutions. The idea is that there should have been many prominent and wealthy individuals, hailing from villages, who are now settled within the urban areas and outside the country and have the determination and benevolence to do something for their motherland. This commitment needs to be channelled for the sustainable development of the villages. Perhaps, it is here the politics of the village may be kept outside the framework and cohesion needs to be achieved. This would certainly supplement the efforts of the Gram Sabha that is the legal and legitimate body elected to care for the development of the village.

The impressive track record of the alumni associations is known worldwide and especially in various areas such as the establishment of educational institutions, starting of new courses in those institutions, setting up of new laboratories and rendering help to their junior student friends. Some old students have donated for the construction of separate blocks and the establishment of state of the art facilities in the hospitals in the healthcare sector. Narendra Modi's concept of Swatch Bharat can be implemented very effectively through this mechanism. One example of how institutions get benefitted by the alumni network is Stanford University. Mr Larry Page and Servey Brin were PhD students of Stanford when they began a research project in 1996. They registered the domain name Google on 15 September 1997 and finally incorporated the company on 4 September 1998. The interesting aspect of the episode is that those alumni got the patent registered in the name of Stanford University and the University is still getting a royalty on that. This can be said to be a classic example of the contribution of great and achieving alumni.

7 Policy Implications

- Integrating technology with business practices can go a long way in yielding desired results.
- The difficulties experienced in the implementation of the project need to be sorted out on a continuous basis. In the case of e-Choupal, Internet appeared to be a bottleneck.
- The experience through the case studies shows that there is a need to have a dialogue with the middlemen, rather than attempting to eliminate them from the system.
- Assured commitment on the part of corporates that they can produce miracles in changing the lives of the rural population. In this case, ITC has been taking every care from the seed development to that of marketing.

  If this model is replicated for a majority of the crops and produce, price discovery for agricultural products would turn out to be real and sustainable.
- There is also a need to rope in as many business establishments as possible to design practices such have been devised by ITC.
- The effort of the governments could also be concurrent and joint with the private sector companies.
- There is also the case for designing alternative models for the development of rural areas with the support of NRIs belonging to the villages who carry abundant love and affection for their native land.

References

Bader, G. (1995). Gandhi on Trusteeship: A Transforming Ethics. *World Business Academy Perspectives, 9*(41), 1.

Bentham, J. (2009). *The History of Utilitarianism*. Stanford: Stanford University.

Bhatia, T. K. (2007). *Advertising and Marketing in Rural India*. Delhi: Palgrave Macmillan.

Bowonder, B., et al. (2002). *Developing a Rural Market E-hub: The Case Study of e-choupal Experience of ITC*. Background Paper Prepared by the Authors for Discussion at the Meetings of Planning Commission of India. Retrieved from www.planningcommission.gov.in

Das, S. (2017). E-NAM Remains Nonstarter. *The Financial Express*.

Gandhi, M. K. (1959). *India of Dreams*. Ahmedabad: Navajivan Publishing House.

Goyal, A. (2010). Information, Direct Access to Farmers, and Rural Market Performance in Central India. *American Economic Journal: Applied Economic, 2*(3), 22–45.

India Philanthropy Report. (2017, March 4). Retrieved from www.bain.com/India-Philantrophy-Report-2017.aspx.

Kalam, A. P. J., & Singh, S. P. (2011). *Target 3-Billion: Innovative Solutions Towards Sustainable Development*. India: Penguin.

Mill, J. S. (1863). *Utilitarianism*. Savill & Edwards Printers.

Pyarelal. (1977). *Mahatma Gandhi on Human Settlements*. Ahmedabad: Navajivan Publishing House.

Rajalakshmi, N. (2017). Why e-NAM Platform Hasn't Taken off Despite all the Fanfare. *The Hindu Business Line*.

Raju, G. N., & Rao, K. V. (2017). *Business Ethics and Corporate Governance*. New Delhi: I.K. International Publishing House.

Rajvanshi, A. K. (2002). Talukas Can Provide Critical Mass for India's Sustainable Development. *Current Science, 82*(6), 632–637.

The Hindu. (2017). The Flaming Fields of Punjab, October 28, 11.

UN. (2015). *The Millennium Development Report 2015*. New York: United Nations.

CHAPTER 3

Skill Development: Role of Industry-Academia Dyadic Collaboration for Sustaining the Construction Supply Chain in Rural India

Sushmera Manikandan, Balan Sundarakani, and Vijay Pereira

1 Introduction

Today managerial decisions are critical when it comes to sustainability aspects of organisational performance and success of their supply chains. Early research (for instance by Mintzberg (1990) discussed supply chain

S. Manikandan
Swiss Business School, Zurich, Switzerland

B. Sundarakani (✉)
Faculty of Business, University of Wollongong, Dubai, United Arab Emirates
e-mail: balansundarakani@uowdubai.ac.ae

V. Pereira
Department of Humanities and Social Sciences, Khalifa University, Abu Dhabi, United Arab Emirates
e-mail: vijay.pereira@port.ac.uk

H. Chahal et al. (eds.), *Sustainable Business Practices for Rural Development*, https://doi.org/10.1007/978-981-13-9298-6_3

management (SCM) as a management philosophy which advocates the integration and coordination of intra-organisational aspects as well as inter-organisational operational and strategic capabilities for the improvement of long-term supply chain performance. Also, C-level executives act as an important facilitator for the implementation of sustainable policies in SCM and the whole academic-industry collaboration spectrum. In this regard, Mulla et al. (2015) suggested that partnership, trust and bonding play an important role in improving the organisation's competitive advantage for greater industry performance. We argue that aspects such as trust and bonding become more important in sustainability in a rural context.

'Academia–Industry' Partnership

Understanding of effective collaboration between academics and industry warrants the need for skills development in the contemporary business world. Thereby, it is argued that the industry should engage not only in research funding and allocation but also in skill development, innovation and entrepreneurship, new technologies, and so on as well as support academic institutions in their common goals (Elliott 2011). Further, the argument that academics should provide incentives to the faculty members who are engaged in research work sponsored by industry in terms of better performance and recognition towards assessment for promotion is also valid. In this regard, Beard (1994) brought in various recommendations for academia-industry collaboration. For example, academics are encouraged to work closely with industrial bodies by undertaking consultancy projects on a regular basis so that they remain entrenched in the real world and ground realities of the market and environment that they operate within. It is in this context that Rizvi (2003) highlights that the academia-industry collaboration becomes vital for sustaining the competitive advantage for both parties.

However, to be successful in this endeavour, practitioners should make sure that they work closely with academics by developing an environment of communication channels, sharing intellectual property rights and data access rights in addition to aiming for Return on Investment (RoI; Ravikrishna 2011). In addition, both parties have to maintain fair and effective policies and procedures for reporting research misconduct as well as other violations of university policies or legal rules and follow prescribed and globally accepted ethical best practices. This could then result in a 'win-win' collaboration, wherein firms can partner and benefit from innovative, scientifically tested and proved methodologies and gain from the expertise and state-of-the-art access to the latest research outcomes by academicians. However, this partnership journey is still at a very nascent stage and, therefore, needs a strong push from both parties (Groznika and Trkman 2012; Lavanya 2013; Mintzberg 1990; Silvestre 2015; Singh and Bhowmick 2015; Tourmen 2009).

2 Literature Review

In an early research, Groznika and Trkman (2012) discussed some key challenges and difficulties of the construction industry such as mitigation risks, business process renovations, trust-based relationships and lack of leadership style and quality, and so on. In addition, the research highlighted some of the risks faced by the construction industry executives such as health, safety, weather, insurance, finance, environment and policies. They also discussed a manager's need to mitigate the various risks of the industry without affecting the firm's profit/bottom line. Another challenge is the risk of speculative projects where builders have to bear the risk in the fluctuating market that will produce lower selling prices than they had anticipated because of the large number of homes that are built speculatively against the market demand which remain unsold (Haenssgen 2018).

On the other hand, Lumineau et al. (2015) revealed that managing suppliers in order to avoid delay and meet the firm's quality standards, maintaining the health and safety measures and quality has been challenging as well. Often projects get delayed due to these reasons that ultimately affect the bottom line of the construction firms. Indeed, lack of trust among the supply chain partners seems to be the biggest constraint for good supply chain coordination and further improvement of supplier production and capacity performance becomes essential. Hence, managers face such challenges in the way of managing trust among investors, developers and contractors across the supply chain network. Recently, Mulla et al. (2015) proposed the role of partnership, commitment, risk sharing, mutual trust and bonding which plays a significant role in improving the organisations' competitive advantage among the supply chain participants in the construction project.

Further, sustainable business practices in this industry have a huge impact in the selection of responsible supply chains such as suppliers and subcontractors that help reduce ecological damage, provide energy efficient designs, sink surplus during construction of materials while also maintaining public health, safety and employees' rights irrespective of gender and ethnicity (Baruah 2010). In addition, community design and their construction affect the social life of the residents and can even lower crime rates as well as advantage all stakeholders, that is, employees, suppliers, buyers and even the government (Li and Lin 2006). Similarly, construction employees have to be paid appropriately for their work in construction projects then be laid off for a few weeks legally/ethically for months until another project starts or they can be given a break during the summer season. Hence, there are challenges as well as opportunities for firms to make Corporate Social Responsibility (CSR) decisions to demonstrate their level of corporate responsibility. Moreover, rural Indian culture has a magnificent effect on creating sustainable business practices which require current businesses to adopt CSR policies and actions (Hofstede 1991).

Having identified these issues, trust is the most important element for an improved supply chain performance and for having long-term supplier/customer partnership (Kim and Choi 2015). In another recent study, Hingley et al. (2015) witnessed how value is formed from a closer buyer and supplier relationships. Nevertheless, one of the major inter-organisational governance mechanisms is a legal contract of trading partners. Since not all companies use the same widely accepted standards, legal contracts enable cooperation where information is shared, as mentioned by Chu and Wang (2012). Therefore, construction industry's challenges are to arrange all these legal contracts to share an understanding of the specific benefits of information sharing between the participating organisations.

3 Challenges of Industry-Academic Synergy

This section illustrates the existing challenges of industry–academia synergy by taking the industry context of construction industry in rural India which is a major driver of the country's economy as more than 70 per cent of India's population lives in rural India (Prahalad and Hammond 2002).

Barriers Due to the Role of Organisational Structure

In general, the construction companies lack the organisational structure that is required to facilitate the industry's academic synergy by establishing functional division. The shortfall exists as these organisations focus only on their individual profit and company's growth and not on the goals of the supply chain (Hofstede 2011; Lavanya 2013; Mintzberg 1990;

Silvestre 2015; Singh and Bhowmick 2015; Tourmen 2009). Hence, there exists some conflict which affects the entire supply chain management. In rural India, most of these developers or sub-developers either have a family-owned organisational structure or no well-developed or established structured organisation (Babu et al. 2017). However, large-scale developers, conglomerates and government housing boards have a well-structured organisational structure. Therefore, in rural cities' development, on-time project execution becomes a challenging task for these family-owned businesses or small enterprises in this industry.

In addition, the organisational structural deficiency of a construction company leads to excessive bureaucracy, a lack of transparency, cultural stratification, political power play, a misalignment of incentives and excessive inter-organisational conflict. Based on the researcher's experience as a manager having worked for a sub-developer, they faced issues in developing appropriate leadership and management styles, balancing operational and strategic pressures, managing workload among team members, achieving Key Performance Indicators (KPIs) based on project performance measures, maintaining staff morale and employee engagement. Anant (2012) indicated in the research that by proper employee training, leadership's succession planning and planning and control of construction operations, construction industries can achieve sustainable supply chain performance in India.

Talent Management

Since the construction industry relies heavily on labour and less on technology, employees are considered to be the greatest assets of this industry. In the case of rural India, the construction industry has been suffering due to out-flux of employees to Gulf countries and Singapore and, therefore, facing a shortage of skilled manpower, technical staff, mason and engineers. Though there are a variety of projects such as education, infrastructure, healthcare and residential projects that form a top priority for Indian governments, shortages in labour, lack of skilled workers and labour policy issues continue to constitute a major challenge due to rising construction material costs. These challenges result in project delays, cancellations and cost overruns (of which the governments are nowadays working towards mitigating risks). The manager is also responsible for taking strategic planning and decisions in order to deliver the projects on time according to our business plan. Also, there exists a lack of standard pay scale among employees of different regions in this industry.

Lack of Quality Excellence

Sourcing of raw materials from listed good quality suppliers is considered to be a big challenge among many off-plan real-estate developments. Also, 'Vasta and Favouritism' play a main role in purchasing of substandard materials, that is, 'I buy from you and you buy from me strategy'. Overall, quality of the product coming from different suppliers may be compromised and reduced substantially. Hence, there is a need to have proper auditing and quality inspection checkpoints at each stage of the supply chain. Therefore, managers are responsible for managing the inventory, suppliers and their products to sustain the safety and quality standards; otherwise these concerns will have a huge impact on the environmental and sustainable aspects of the company's reputation and cash flow.

Areas of Uncertainty

There exists some degree of demand uncertainty in material costs due to the fluctuating real-estate market which affects the overall Indian economy. In the case of such supplier and developer the industry is moderately regulated as there has been lack of standard and government policies yet. India's infrastructure investment is a very important driver for the economy, so the government should be associated with other forms of financing such as the PPP (public-private partnership) projects that require initiative at critical fluctuating market situations. Hence, the increase in the country's population leads to surging demand for residential and commercial housing. It also outweighs supply; hence the increase in construction of infrastructure projects, shopping malls, hospitals and schools becomes a big challenge, and both go hand in hand according to a growing population.

Rising Construction Costs

Another challenge increasingly faced by firms is the entry of foreign workers in the country. The rising costs of raw materials such as cement, steel and fit-out furniture accounts for 30 per cent of construction costs. In order to minimise the construction cost proactively, proper project management of mapping the constructional activities of various stages becomes a pre-requisite.

Technology: Threats/IT Issues

Although India is one of the countries with a large IT industry skill set, outsourcing of projects to technology- and manufacturing-related jobs created a significant negative pressure for people with low technology skills which in turn threatened the construction industry. Alternatively, cyber-attacks are a constant risk which results in commercial information loss, exposure of confidential data which adversely impacts a company's reputation and ability to continue the business. Therefore, managers need to prepare for possible disruptions and various risk mitigation strategies as well as work closely with academics for contingency planning.

Market Growth Research

India's research and development is considered as the crucial factor behind the development of effective and efficient construction processes, advanced technologies, procurement routes and managerial structures. In addition, R&D activities always create the opportunity to be competitive in the global market for the success of construction organisations. According to Perkmann and Walsh (2007), this established a synergistic relationship and managing research partnerships takes time to enhance the entire teaching-learning process. Moreover, another challenge is the existence of a different hierarchy of processes and policies for allocation of funds and securing funding to different nationalities and cultures for collaborative research with universities and researchers of the country (Kaklauskas et al. 2018). Similarly, another challenge is identifying emerging research trends and building relationships to research excellence as strategic initiatives and presenting their challenges to the community.

Ethical Challenges

Fundamental ethical academics are a primary concern because academic deceit affects both the integrity of the curriculum and overall reputation of the educational institution. Hence, higher level of trust should be maintained to determine a firm's success and sustainability with academic collaboration. Next, the biggest ethical issue is a fiscal conflict of interest which biases decision-making and weakens adherence to professional, ethical or legal international standards. Unethical behaviour by the construction industry parties impacts the quality of project performance categorised by Rahman et al. (2007).

By identifying these challenges pertaining to the Indian rural construction industry, we now apply the Mintzberg (1973) theory to assess the different roles that managers should undertake towards an effective organisational execution and business excellence.

4 Henry Mintzberg Management Roles

According to McCrimmon, 'Management is the process of allocating resources in order to achieve goals in maintaining a sustainable beneficial environment where people could fulfill their obligations and goals through trust based long term collaborative approach'. Concurrently, Mintzberg (1973) explained that a wide variety of manager's roles in an organisation are classified into interpersonal, informational and decisional roles.

Type 1: Interpersonal Roles

This role covers the interactions and relationships among the team members and other constituents in a firm in order to achieve organisational goals.

Figurehead

Firstly, not many industry practitioners are interested in sharing their experience with academics. Secondly, if at all some are interested, a few of them indeed look for some form of return commitments. Unfortunately, most academic institutions are non-profit organisations that hold minimal reserve funding for industry collaborations. Moreover, it's difficult to get an appointment from executives due to different organisational hierarchical structures of decision making.

Leader

A leader has to capitalise in developing a team's management skills and also reduce costs while improving customer service and support the expansion of new market. Thereby, both of them can benefit from the academic conversation by building new knowledge and disseminate this new knowledge for improvement of industry through extensive engagement with corporate representatives. Hence, the biggest strategic issues are the supply chain transformation, leadership, sustainability and innovation. However, the partnership between supply chain faculty and top executives in the construction industry leads to a firm's mission of increased trust-based collaboration with different tiers of suppliers, greater emphasis on quality, more innovative product design and increased cost savings and revenues. Also, maintaining leadership qualities such as communication skills, decision making, integrity, organisational culture, health and safety, business ethics and sustainability towards a higher level of productivity.

Liaison

A liaison is mainly responsible for managing a fully outsourced supply chain for performing activities related to the endowment and supply of construction materials, inventory management and logistics. In addition, they also maintain contacts through social networking with all their customers to satisfy their needs.

Type 2 Informational Roles

This role consists of sharing, collecting and disseminating information to successfully achieve organisational goals.

Monitor

A monitor determines an accurate way of checking the quality of construction materials by considering the accuracy, quality control, level of productivity, meeting targets, negotiations, supplier performance, health and safety and process of a construction project. Construction performance monitoring is at the heart of the performance management processes, which allows project participants to identify any bottleneck in a supply chain across organisational boundaries for supply chain process evaluation and improvement. Based on the employee's feedback, competitor's sales pattern, survey and questionnaire, academicians have to stay up to date in their research and sync it to recent industry standards and deviations occurring in both the business environments.

Disseminator

The disseminator conveys the appropriate information gained from both internal and external sources. For instance, organisations which are proactive tend to implement such sustainable practices at an earlier stage to disseminate the best practice to the industry and the followers. However, knowledge dissemination and learning can happen at a different pace through supply chain integration and collaboration (Sakthivel and Joddar 2006). Construction management graduates need to possess three essential skills namely practical experience, management tools and techniques and interpersonal skills. Information gathered as a monitor from different sources should be transferred to higher management for authorisation.

Spokesperson
The spokesperson transfers the organisation's information to all people. Then the coordinated decision should be taken with the manager's approval, a construction industry expert and the supervisor and a unit or departmental expert.

Type 3 Decision Roles

This role involves collecting and gathering the information and resources, which allows a manager to take effective decisional roles and responsibilities.

Entrepreneur
Generally, an entrepreneur is someone who is a motivator, designer and also encourages organisational change and innovation. In addition, different roles encourage managers to create enhancement projects and work to delegate, to authorise and manage team members in the development process of increasing living standards. As the construction industry requires low tech projects, entrepreneurship is considered as a crucial role for sustaining and improving the performance of projects in terms of innovativeness, risk-taking and competitive aggressiveness.

Disturbance Handler
As a manager, corrective action should be taken when the organisation faces unexpected disturbances and financial crises. For instance, corrective action in response to unforeseen problems such as strikes, financial difficulties, uncertainty, forecasting or a change in government policy. In such situations, both practitioners and academicians can plan a new strategy together and provide leadership support to each other.

Resource Allocator
A resource allocator must be responsible for the distribution of the organisation's resources effectively, at the same time they should maintain accountability. For instance resources such as money, equipment, staff members and raw materials have to be allocated efficiently to the working team members.

Negotiator
Negotiation plays a significant role between a subcontractor and the main contractor, a relationship that is primarily based on achieving the lowest price and the underlying risk and stakes of the price. Both the academic and industry partners are expected to have good negotiation skills with proper understanding of industry standards. Negotiations should have been carried out through external parties for contract with suppliers, working terms and conditions with the trade union and environmental impacts (Peterson et al. 2004). In general, negotiations with different suppliers and contract agreements are the biggest challenges of the rural Indian construction industry because of the Asian culture. Moreover, negotiation mostly occurs in material cost, lead time, international quality standards, payment terms, warranties, merchandise return policy, and so on. Katz (1974) has identified three basic types of managerial skills for practising managers such as technical, human and conceptual skills.

Overall, the rural Indian workforce of the construction industry is facing lack of technological know-how and they need skill enhancement in terms of its way of operation. Having discussed the various challenges, the rural population and its mobile technology adoption and penetration seemed to be phenomenal in recent years as reported by Ravikrishna (2011) and Haenssgen (2018). However, there seemed to be a lack of collaborative working culture in the industry.

5 Conclusion

Today's managers experience many supply chain challenges which pose a high threat to the construction industry and their project's success. Some of the identified issues include workforce safety, time constraints, quality of the materials and industry standards. On the flipside, some of the non-construction challenges that have been identified are the lack of government regulations, industry legal issues, environmental concerns and socio-political pressures. However, favouritism and corruption are the main drawbacks of India's construction industry which result from lack of transparency with contractors and accounting enough funds, proper budgeting which affect the industry and its productivity. Moreover, academic-industry challenges have been discussed through Mintzberg's managerial role in which management and co-ordination of projects result from efficient cost-effective planning, time management, communication, progress monitoring strategies, administrative issues which should be transparent to all supply chain parties for the success of Indian construction projects. But in today's competitive world, managers should understand and navigate across the previously discussed challenges and transform the challenges to opportunities.

References

Anant, D. (2012). Supply Chain Management Dimensions, Supply Chain Performance and Organizational Performance: An Integrated Framework. *International Journal of Business and Management, 7*(8), 2–10.

Babu, G. S., Saride, S., & Basha, B. M. (Eds.). (2017). *Sustainability Issues in Civil Engineering* (Springer Transactions in Civil and Environmental Engineering) (Vol. XV, p. 367). Singapore: Springer.

Baruah, B. (2010). Gender and Globalization: Opportunities and Constraints Faced by Women in the Construction Industry in India. *Labor Studies Journal, 35*(2), 198–221.

Beard, C. (1994). Educating the Star Fleet caption-making Business Schools More Relevant to Their Stakeholders. *Working Paper*. University of Paisley.

Chu, Z., & Wang, Q. (2012). Drivers of Relationship Quality in Logistics Outsourcing. *Journal of Supply Chain Management, 48*(3), 78–96.

Elliott, K. (2011). *Is a Little Pollution Good for You?* New York: Oxford University Press.

Groznika, A., & Trkman, P. (2012). Current Issues and Challenges of Supply Chain Management. *Economic research-Ekonomska istraživanja, 25*(4), 1101–1112.

Haenssgen, M. J. (2018). The Struggle for Digital Inclusion: Phones, Healthcare, and Marginalisation in Rural India. *World Development, 104*, 358–374.

Hingley, M., Lindgreen, A., & Grant, D. B. (2015). Intermediaries in Power-laden Retail Supply Chains: An Opportunity to Improve Buyer-Supplier Relationships and Collaboration. *Industrial Marketing Management, 50*, 78–84.

Hofstede, G. (1991). *Cultures and Organizations: Software of the Mind*. London, UK: McGraw-Hill.

Hofstede, G. (2011). Dimensionalizing Cultures: The Hofstede Model in Context. *Online Readings in Psychology and Culture, 2*(1), 12–23.

Kaklauskas, A., Banaitis, A., Ferreira, F. A. F., Ferreira, J. J. M., Amaratunga, D., Lepkova, N., Ubartė, I., & Banaitienė, N. (2018). An Evaluation System for University–Industry Partnership Sustainability: Enhancing Options for Entrepreneurial Universities. *Sustainability, 10*, 119.

Katz, R. (1974). Managerial Skills. *Harvard Business Review, 12*, 23–27.

Kim, Y., & Choi, T. (2015). Deep, Sticky, Transient, and Gracious: An Expanded Buyer-Supplier Relationship Typology. *Journal of Supply Chain Management, 51*(3), 61–86.

Lavanya, B. (2013). A Study on Perceived Benefits from Modes of Industry-Academia Interface'. *Journal of Research in Management & Technology, 2*, 1–8.

Li, S., & Lin, B. (2006). Accessing Information Sharing and Information Quality in Supply Chain Management. *Decision Support Systems, 42*(3), 1641–1656.

Lumineau, M., Stephanie, E., & Handley, S. (2015). Inter-organizational Conflicts: Research Overview, Challenges, and Opportunities. *Journal of Strategic Contracting and Negotiation, 1*(1), 42–64.

Mintzberg, H. (1973). *The Nature of Managerial Work*. New York: Harper and Row.

Mintzberg, H. (1990). The Manager's Job: Folklore and Fact. *Harvard Business Review, 68*(2), 163–176.

Mulla, A., Gupta, A. K., & Desai, D. B. (2015). Novelty Journals Supply Chain Management: Effective Tool in Construction Industry. *International Journal of Novel Research in Engineering and Science, 2*(1), 35–40.

Perkmann, M., & Walsh, K. (2007). University-Industry Relationships and Open Innovation: Towards a Research Agenda. *International Journal of Management Reviews, 9*(4), 259–280.

Peterson, T., David, D., & Fleet, V. (2004). The Ongoing Legacy of R.L. Katz: An Updated Typology of Management Skills. *Journal of Management Decision, 42*(10), 1297–1308.

Prahalad, C. K., & Hammond, A. (2002). Serving the World Poor, Profitability. *Harvard Business Review, 80*(9), 4–12.

Rahman, H. A., Karim, S. B. A., Danuri, M. S. M., Berawi, M. A., & Yap, X.W. (2007). Does Professional Ethic Affects Construction Quality? In *Quantity Surveying International Conference*, Kuala Lumpur, Malaysia.

Ravikrishna, R. V. (2011). Sustainable Energy for Rural India. *Journal of Religion and Science, 46*(4), 942–956.

Rizvi. 2003. An Assessment of Academia-industry Partnerships Through Distance-learning Technology Utilization. Pepperdine University, Education, ProQuest, 2008, 153.

Sakthivel, S., & Joddar, P. (2006). Unorganised Sector Workforce in India: Trends, Patterns and Social Security Coverage. *Economic and Political Weekly, 41*, 2107–2114.

Silvestre, B. S. (2015). Sustainable Supply Chain Management in Emerging Economies: Environmental Turbulence, Institutional Voids and Sustainability Trajectories. *International Journal of Production Economics, 167*, 156–169.
Singh, S., & Bhowmick, B. (2015). An Exploratory Study for Conceptualization of Rural Innovation in Indian Context. *Procedia: Social and Behavioral Sciences, 207*, 807–815.
Tourmen, C. (2009). Evaluators' Decision Making: The Relationship between Theory, Practice and Experience. *American Journal of Evaluation, 30*(1), 7–30.

CHAPTER 4

Understanding and Analysing the Role of Knowledge, Attitude and Practices for Prevention of Diabetes

Sandeep Narula and Sunita Nigam

1 Introduction

Presently diabetes, in India, is an epidemic and India is considered as the diabetic capital (Gupta and Gupta 2009). According to the World Health Organization (WHO) and the International Diabetes Federation (IDF), diabetes has become the primary global healthcare challenge. Type 2 Diabetes Mellitus (T2DM) is widely recognised as a non-communicable disease (NCD) and has reached and has the potential of becoming an epidemic in India and if complete awareness and education are not imparted to the Indian population then certainly it can take a heavy toll on the Indian society. It is a silent killer and essentially co-factors with other diseases too and thus the level of morbidity and mortality is much high. Hence, it can pose a significant burden on the government and society

S. Narula (✉)
IIHMR University, Jaipur, India
e-mail: sandeep@iihmr.edu.in

S. Nigam
School of Health and Hospital Management, IIHMR University, Jaipur, India
e-mail: sunita@iihmr.edu.in

H. Chahal et al. (eds.), *Sustainable Business Practices for Rural Development*, https://doi.org/10.1007/978-981-13-9298-6_4

(Narula 2016). It is estimated that in 2000, there were approximately 150 million individuals with this disease and that this number is likely to double by 2025 (King et al. 1998). Further, since T2DM is genetic and catastrophic, if left unchecked can severely affect the younger population, which will be more lethal and dangerous to a young nation. Moreover, the younger population is more prone to economic boom and a sedentary lifestyle and is certainly more vulnerable to T2DM (Narula 2016).

Very high levels of changes in the T2DM have been observed when there is a big change in the lifestyle, eating patterns, including the type of diet consumed, particularly from local or indigenous habits to western or junk food (Hetzel and McMichael 1987). Changing disease rates are almost certainly explained by changes in several dietary factors as well as by changes in other lifestyle related factors, notably a reduction in physical activity (Narula 2016).

2 Literature Review

Fuentes (2000) believes that NCD is a more common cause of death than infectious diseases. There is enough literature supporting the fact that unhealthy lifestyle and faulty dietary habits increase the incidence and prevalence of NCDs. Although there are lot of resources on 'What is to be done' for the prevention of NCDs such as diabetes, the major challenge is 'How should it be done'.

The current chapter highlights the prevailing knowledge and attitude about diabetes and attempts to validate the findings to figure out what practices need to be altered to bring about the right lifestyle change for the prevention of diabetes. This has also been highlighted by Nissinen et al. (2001).

There is sufficient literature available citing that there was a positive intervention effect; proportion varying from 83% in case of family only (10/12), 87% in case of family and community (7/8) and 76% that involved school only (13/17) as per Saraf (2012). McQueen (2013) supports two types of approach for the prevention of NCDs: 'Life course approach' and 'population approach'. The 'life course approach' aims at the prevention and control of NCDs in the population starting with maternal health, prenatal nutrition, pregnancy outcomes, proper feeding practices in infancy and child and adolescent health of children at school, youngsters at college, followed by interventions targeting adults to encourage healthy diet, regular physical activity and avoidance of tobacco from youth into old age. Darnton-Hill et al. (2004) also believe that people with this disease will need cost-effective medical interventions, whereas the 'population approach' aims at reducing the risk factors in the community. Since this approach works on the continuum of risk factors, the mass change also results in mass benefits. Under this approach, the individuals with high risk factors are targeted for interventions to reduce risk factor levels. These interventions involve increasing knowledge, awareness about the disease and its associated factors and bringing improvement in lifestyle practices (Singh and Reddy 2011). Obesity and being overweight are best controlled by adopting healthy diet practices such as avoiding excess calorie food intake and opting for more high-fibre fruits and vegetables as these have shown favourable effects on body lipid levels and have been supported in the studies of WHO (2004), Katan et al. (1995) and Penny and Etherton (1999).

Similarly, increased physical activity reduces obesity, cholesterol and keeps a check on the weight and the interventions can be suggested at the family/community/national/regional level as mentioned in the work by Matsudo and Matsudo (2004). Moreover, Yadav and Krishnan (2008) have mentioned that physical inactivity and sedentariness are common in both genders in India, as it has been observed that physical activity is also related to occupation and spare time activity. Purran et al. (1995) has cited that community-based intervention has also produced remarkable results over a long period such as the one in North Karelia, Finland; it significantly lowered the risk factors and mortality rate. Prevention, Awareness, Counselling, Evaluation (PACE) community-based programme carried out in Chennai has shown significant improvement in the lifestyle, dietary habits, physical activity and prevention and control of diabetes as shown by Somannavar et al. (2008) in their research work.

For prevention and management programme which included residents, peer educators and village leaders/sarpanchs, the intervention results were highly impressive in terms of blood fasting glucose levels across healthy, high-risk and T2DM individuals as observed by Prabhakaran et al. (2009). Very few studies have been done in India for the determinants of health behaviours. A study carried out in South India by Murugesan et al. (2007) found that there is very low level of knowledge in context of the advantages of a healthy lifestyle, causes of T2DM and then prevention and management of this disease especially in women, unskilled labours/workers and also among the population with low education level. According to Joshi (2008) and Wilson (2010), diabetes management is inadequate not only due to poor medical control but also due to a few other factors such as shy behaviour and reluctance to share information of the illness even with their friends and colleagues/peers.

3 Conceptual Framework

Keeping in mind the unhealthy behaviour and practices, there is a need to adopt a framework which not only addresses an individual's unhealthy behaviour but also enables people, individually and collectively, to increase control over their health and its determinants and thereby promote and sustain good health. Promotion of health relies strongly on mediation and advocacy for these 'enabling environments' that are very much essential for the prevention and control of NCDs.

Hence, the following disease continuum helps us to identify the different levels in which the approaches to health promotion can be applied as shown by Dobe (2012) (Fig. 4.1).

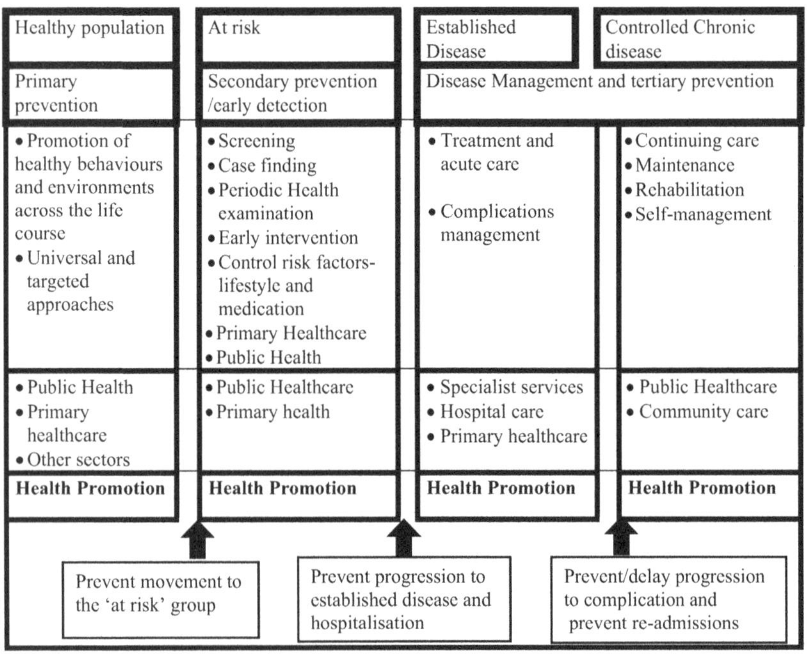

Fig. 4.1 Health promotion at different levels. (Source: IJPH)

4 Problem Statement and Objectives

It is estimated that the prevalence of diabetes in rural populations is one-quarter of the urban population for India and other Indian sub-continent countries such as Bangladesh, Nepal, Bhutan and Sri Lanka (Seema and Cornwall 2014; V. Mohan 2011).

NCDs contribute to around 5.87 million deaths in India. Four major types of NCDs are cardiovascular, cancer, respiratory and diabetes that account for the largest contribution of morbidity and mortality. Major behavioural risk factors are unhealthy diet, physical inactivity, obesity, high blood pressure, high blood glucose and high blood total cholesterol levels (WHO 2014).

Since NCDs are making a disproportionate impact on people at younger ages causing premature loss of life, there is a need to check these unhealthy practices and lifestyle trends prevailing in the community as has been suggested by Mendis (2010).

Keeping in view the point discussed, the following objectives were framed for the proposed survey.

The specific objectives of the project evaluation are as follows:

1. To evaluate the status of the target population (beneficiaries under the project area) on:
   - (a) Knowledge of diabetes, risk factors, symptoms.
   - (b) Health-seeking practices in terms of screening for diabetes.
   - (c) Current load of diabetes and other NCDs in terms of treatment-seeking.
   - (d) Attitude of the people to help diabetes-affected people.
   - (e) Local food habits.
2. To measure the degree of changes in knowledge, attitude and practices (KAP) on diabetes community care and support projects among the target community before and after the implementation of the project.

5 Research Methodology

Project Background of Baseline Survey

This research is an end line survey and the baseline survey was carried out in Jodhpur in December 2012 as a complementary initiative to support the rolling out of The National Programme for Prevention and Control of

Cancer, Diabetes, Cardiovascular Diseases and Stroke (NPCDCS). The NPCDCS is integrated with National Health Mission through the involvement of ASHA workers. The main objective of the diabetes community care and support project is to enhance community education on diabetes, community mobilisation for diabetes testing and community sensitisation to minimise the risk factors of diabetes. The project was supported by the Bristol-Myers Squibb Foundation and completed its tenure on 31 May 2015.

End line Survey

To carry out the end line survey, the team carried out secondary data research of the documents related to the baseline project such as proposal, operational guideline, baseline-survey report, quarterly performance reports, reports submitted to donors, project records, annual report of Humana People to People India (HPPI), secondary report and data and others to develop the study instruments (quantitative and qualitative), process indicators and guidelines for the end line survey.

Sample Size

The sample size was calculated as follows:

$$\mathrm{SS} = \frac{Z^2 \times (p) \times (1-p)}{c^2},$$

where Z = Z value (e.g. 1.96 for 95% confidence level); p = prevalence 0.38 diabetes awareness level in baseline (percentage picking a choice, expressed as decimal: 0.38 is used for sample size needed) and c = confidence interval, expressed as decimal (e.g. 0.05 = ±5).

For quantitative evaluation of KAP with the household (HH) members (considering the prevalence of awareness in Mandore block in Jodhpur District of Rajasthan general, 95% of confidence level and confidence interval 0.05), the sample size calculated was 363. Considering sample loss/sampling error we added another 10% in our sample size hence the total sample for the study was 400. Out of the total 74 villages in baseline coverage, 25% were studied, that is, 20 villages and 20 samples from each village (20 × 20 = 400 sample population) was covered, which gave us a good scientific estimate about the entire population. The sample villages were selected through the Population Proportion to Sample (PPS) method and selection of HH followed the simple random sampling process from the listing of HH in the village.

Study Instrument

The questionnaire was designed using the baseline survey secondary data, as in the end line survey, the same indicators were supposed to be used and measured. The questionnaire was divided in two parts: Section A and Section B.

Section A comprised demographic information including age, number of family members, education, income, dietary habits of households and so on. This section was used at household level.

Section B comprised KAP: Knowledge, Attitudes and Practices pertaining to T2DM. The 20 questions in Section B were developed using NPCDCS guidelines. The data in Section B questionnaire was obtained by one male and female respondent from the households.

Since it is an end line survey, the primary objective was to study and assess the value of the indicators; hence, no hypothesis was required for this.

6 Results

For Section A of the questionnaire, in total, 20 villages were identified for the survey and from each village 10 HHs were selected for obtaining responses. Overall, 207 HHs participated in the survey.

For Section B responses, two respondents, one male and one female were identified to obtain the responses for KAP on T2DM.

Section A Analysis

The end line survey covered total 1618 respondents as can be observed in Table 4.1. The table reveals the demographic classification of the sample, which highlights that male earning members are 23% of the sample and approximately half of the population is in the age bracket of 18–60 years.

Table 4.2 denotes that 47% of the members are in the age bracket of 18–50 years and contribute 77% to the earnings. Nearly 42% of the HH comprise non-agricultural labour, followed by 13% agriculture labour,

Table 4.1 Demographic breakdown of the respondents

Particulars				*Earning members*		
Age	*Male*	*Female*	*Total*	*Male*	*Female*	*Total*
Number of HH members above 60+ yrs	83	75	158	24	2	26
Number of HH members between 50–60 yrs	80	67	147	50	4	54
Number of HH members between 40–50 yrs	102	103	205	88	7	95
Number of HH members between 30–40 yrs	134	134	268	118	4	122
Number of HH members between 18–30 yrs	156	129	285	78	2	80
Number of HH members below 18 yrs	301	254	555	10	0	10
Total members in HH	856	762	1618	368	19	387
%	*53%*	*47%*		*23%*	*1%*	*24%*

Table 4.2 Profession-wise details of the respondents

Particulars	*Frequency*	*Percentage*
Non-agricultural labour	86	41.5
Domestic servant	2	1.0
Agricultural labourer	26	12.6
Cultivator	16	7.7
Petty business/small shop owner	24	11.6
Small artisan in HH/cottage industry	12	5.8
Self-employed	10	4.8
Employed (government or private)	22	10.6
Large/medium business	4	1.9
Other (specify)	5	2.4
Total	207	100.0

clearly showing the dominance of labour class in the strata. The average monthly income of the family is between Rs. 20,000 and 40,000 and that constitutes 82% of the population, signifying a typical middle-class income distribution (Fig. 4.2).

The dietary habits of the sample HH is given in Table 4.3. Of the total sample, 81% of the families are vegetarian and 19% are non-vegetarian. During the survey, it was established that nearly 66% people have meals three times a day, and in addition, nearly 40% of the population consumes

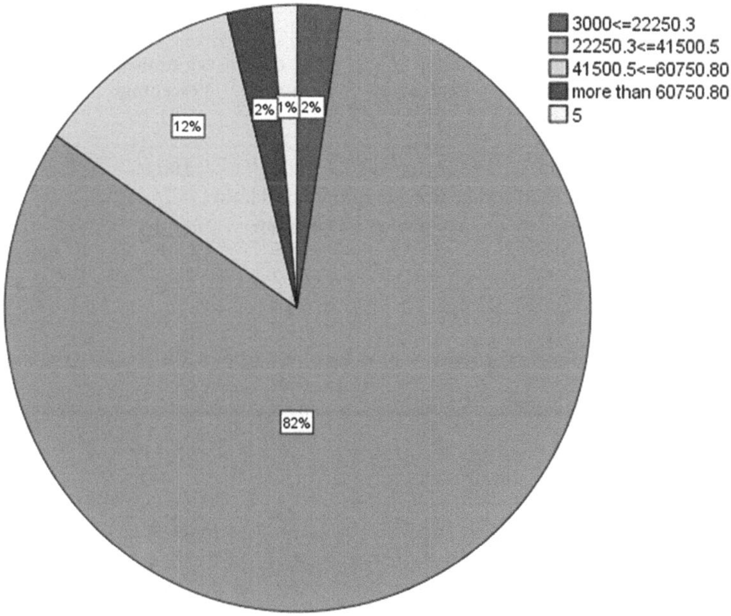

Fig. 4.2 Average monthly income of the family (binned). (Source: Primary data collected by the authors)

'junk food'—frequency varying between several times to once in a month. From Table 4.3 data we can also deduce that only 41 of the respondents have breakfast, challenging their response of three meals a day. Moreover, among them, they also consume a good quantity of 'junk food' which possesses high amounts of calories, salt, oil and other fatty acids, which are certainly not good for health. Among the beverages, 98% people consume tea as the preferred beverage.

The most significant finding of the research is the pattern of alcohol consumption, which has increased considerably as compared to the baseline. As evident from Table 4.4, a large proportion of the survey respondents informed us that none of their household members consume alcohol (84.5%). But the proportion of household members habituated to alcohol consumption and smoking or both has increased consistently from baseline to end line. Moreover, alcohol is prone to cause an increase in calories (Traversy and Jean-Philippe 2015). In a report by the Department of

Table 4.3 Dietary habits of respondents

No. of vegetarian household/families			**No. of non-veg household/families**	
	Frequency	**Percentage**	**Frequency**	**Percentage**
No	39	18.9	168	81.16
Yes	168	81.1	39	18.84
Total	207	100.0	207	100.0
Average frequency of food-intake per day in the household				
	Frequency	**Percentage**		
2 times	59	28.5		
3 times	137	66.2		
4 times	9	4.3		
more than 4 times	2	1.0		
Do you/family members eat snacks or junk food and what is the frequency				
Yes	135	65.2		
No	72	34.8		
Total	207	100.0		
Several times a week	9	4.3		
Once a week	23	11.1		
Once a fortnight	2	1.0		
Once in a month	20	9.7		
Once in several months	4	1.9		
Rarely	77	37.7		
Total	135			

Do you/family members have Breakfast			**Beverages**	
	Frequency	**Percentage**		**Frequency**
Yes	41	19.8	Yes	204
No	166	80.2	No	3
Total	207	100.0	Total	207
If yes, what			If yes, what	
Stuff fried bread (Kachori)	6	2.9	Tea	200
Stuffed potato dough (Samosa)	4	1.9	Coffee	5
Sweet dough sugar swirl (Jalebi)	1	.5	Cold drink	2
Pulse round cake (Pakora)	4	1.9	Squash	0
Chilly round cake (Mirchi Bada)	2	1.0	Milk	24
Bread	7	3.4	Others	1
Dehusked flattened rice (Poha)	5	2.4		
Wheat flour bread (Parantha/Roti)	13	6.3		
Others	5	2.4		

Table 4.4 Substance habits consumption

	Baseline	*End-line*
Proportion of households where members are habituated to alcohol consumption	5.6	15.5
Proportion of households where members are habituated to smoking	15.8	35.3
Proportion of households where members are habituated to both alcohol and smoking	1.5	9.7
Alcohol consumption		
One member		12.6
Two members		1.9
Three and above		1.0
All members		0.0
None		84.5
Smoking habit		
One member		30.4
Two members		3.9
Three and above		1.0
All members		0
None		64.7

Health and Human Services, it is said that if a person has diabetes and if he/she smokes, then the person is more likely to have serious health problems from diabetes (U.S. Department of Health and Human Services 2010).

Disease History of the Household

The founder member of the Delhi Diabetes Forum says that the prevalence of diabetes increases with a family history of diabetes. The risk of a child developing diabetes with a parental history increases above 50%. Questions were asked from the respondents in context to the family history of diabetes, hypertension and diabetes related complications. A comparison made between the baseline and end line is shown in Table 4.5 which reveals a high incidence of diabetes among the first-degree relatives. As compared to the baseline, the proportion increased significantly in the end line cases, hence again highlighting the need to improve the KAP of the people. Only 2% HHs have a history of diabetes and 5% respondents had any family history of hypertension. Only 7% HHs had a history of both diabetes and hypertension. The following table shows a comparison of the diabetes disease baseline versus end line across age groups of the households.

Table 4.5 Household members suffering from various diseases baseline and end line

	Baseline	*End-line*
Characteristics		
Proportion of households with history of diabetes	0.97	1.9
Proportion of households with history of hypertension	0.58	4.8
Proportion of households with history of both diabetes and hypertension	1.55	6.8
Proportion of households where anyone died in the household due to diabetes or diabetes related complications	0.05	0.5
Proportion of households with a history of diabetes (age group)		
Households < 30 yrs	7.6	0.0
Household between 3045 yrs	50.4	52.2
Households > 45 yrs	42.0	47.8
Percentage of MALE	36.0	49.8
Percentage of FEMALE	64.0	50.2

Note: This table is based on the HH data and not on individual responses. HH total 1618 is a proportion not frequency and not per cent.

Table 4.5 signifies that people in the age group of 30–45 years and 45 years are in the high-risk zone, hence this served as the basis of the concerted efforts for the project.

Diabetes Association-Statistical Significance

Before carrying out the level and intensity of KAP, statistical tests were performed to assess the significance of level of diabetes with:

- Age
- Gender
- Occupation using Chi-Square test of independence.

Followed by the two-way Analysis of Variance (ANOVA), testing was done to measure the significance of diabetes with body weight and smoking. All the variables considered earlier are used in KAP for community engagement interventions. A sample of 200 was taken which included 46 diabetic and 154 non-diabetics to assess the impact of KAP.

Chi-Square Test of Independence

$H_0 1$: There is no relationship between diabetes and Age, Gender and Occupation

Table 4.6 Chi-square test of independence among the variables

Cross tabulation—gender and occupation				
		Do you have diabetes?		**Total**
		No	**Yes**	
Gender	Male	92	23	115
	Female	62	23	85
Total		154	46	200
Is your occupation?	Inactive	32	7	39
	Light work	92	29	121
	Heavy work	30	10	40
Total		154	46	200
Chi-Square tests—Age				
	Value	**df**	**Asymp. Sig. (2-sided)**	
Pearson Chi-Square	19.962[a]	19	0.397	
Likelihood ratio	19.751	19	0.410	
Linear-by-linear association	0.068	1	0.794	
No. of Valid Cases	200			
Chi-Square tests—Gender				
Pearson Chi-Square	1.375[a]	1	0.241	
Likelihood Ratio	1.365	1	0.243	
Linear-by-linear Association	1.368	1	0.242	
Chi-Square tests—Occupation				
Pearson Chi-Square	0.716[a]	2	0.699	
Likelihood Ratio	0.747	2	0.688	
Linear-by-linear association	0.546	1	0.460	

[a]Zero cells (0.0%) have expected count less than 5. The minimum expected count is 8.97

As it is evident from the results of Table 4.6, the Chi-Square test proved that there is no relationship between diabetes and age, gender and occupation, as it is independent of age, gender and occupation, $H_0 1$ (null hypothesis) was accepted. Hence, the KAP needs to be imparted to all the persons irrespective of age, gender and occupation.

Two-way ANOVA

Two–way ANOVA was also performed to assess the effect of body weight (dependent variable) on diabetes (independent variable with two levels—Yes/No) and smoking (independent variable with two levels—Yes/No) and the interaction effect between the smoking and diabetes. Each of the tests was studied at 5% significance level ($\alpha = 0.05$).

So, the three sets of hypotheses for the two-way ANOVA are as follows:

$H_0 2a$: The population means of factor smoking (smokers and non-smokers) are equal.

Table 4.7 Two-way ANOVA assumption test of equality of variances

Levene's test of equality of error variances			
Dependent variable: Your_weight			
F	*df1*	*df2*	*Sig.*
1.055	3	196	0.370

Tests the null hypothesis that the error variance of the dependent variable is equal across groups

Design: Intercept + DV_Do_u_hv_diabetes + Do_u_smoke + DV_Do_u_hv_diabetes × Do_u_smoke

$H_0$2b: The population means of factor diabetes (diabetic vs. non-diabetic) are equal.

$H_0$2c: There is no interaction between smoking and diabetes.

All three null hypotheses were tested, and the results were found to be statistically significant along with the assumption that variances of the population must be equal was also true as shown in Table 4.7.

Smoking and diabetes have a significant effect on body weight and they can be considered as comorbid factors for diabetes. Hence, patients need to check their body weight and smoking to stay away from diabetes.

As we can see from Table 4.8 that both the test of effect are significant, as *p*-value is less than 0.05 and also there was significant onset of diabetes by smoking interaction effect on body weight of the patients with F (1196) = 8.005 with $p = 0.005$ and η^2 (partial eta square) = 0.039; the interaction is also evident from the following graph as the lines are non-parallel (as parallel lines indicate no interaction) (Fig. 4.3).

KAP Level Information

For assessing the KAP of the disease in the community, 408 respondents were considered for the survey. When asked, have you heard about diabetes, 96.3% of the respondents said 'yes'. When asked what they meant by diabetes, 96% people replied that it's a disease. When the respondents were asked from where they have heard about the disease, their responses varied and these have been elaborated in Table 4.9.

Table 4.8 Test of effect of both the factors and interaction effect

Tests of between-subjects effects

Dependent variable: Your_weight

Source	*Type III sum of squares*	*df*	*Mean square*	F	*Sig.*	*Partial Eta squared*
Corrected model	3306.319[a]	3	1102.106	22.154	0.000	0.253
Intercept	484,256.112	1	484,256.112	9734.389	0.000	0.980
DV_Do_u_hv_diabetes	1287.167	1	1287.167	25.874	0.000	0.117
Do_u_smoke	201.041	1	201.041	4.041	0.046	0.020
DV_Do_u_hv_diabetes × Do_u_smoke	398.216	1	398.216	8.005	0.005	0.039
Error	9750.401	196	49.747			
Total	1,047,554.000	200				
Corrected total	13,056.720	199				

[a] R squared = 0.253 (adjusted R squared = 0.242)

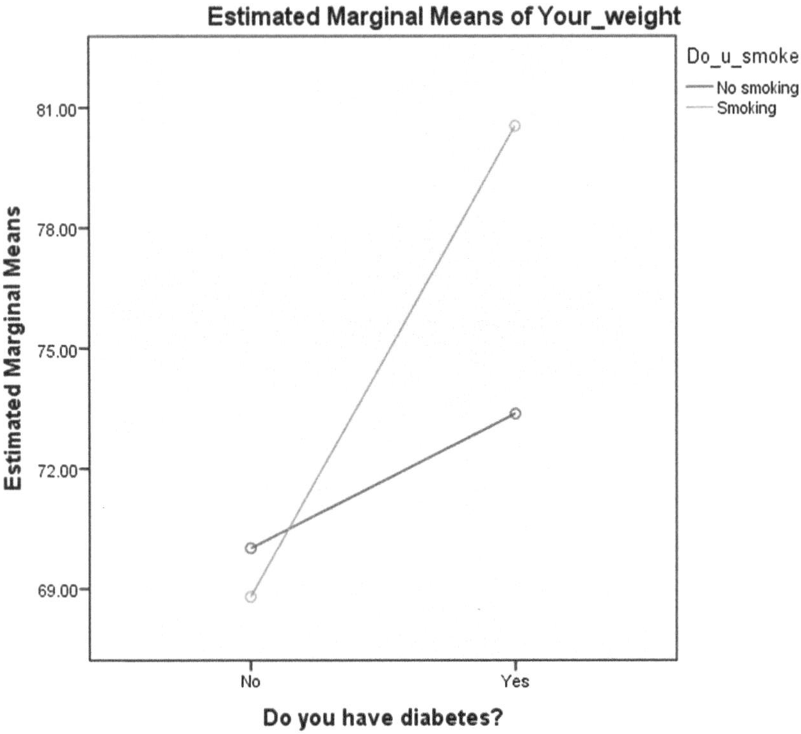

Fig. 4.3 Interaction effects of the factors

Diabetes Awareness

When respondents were asked, if they knew the symptoms of diabetes, 58% respondents said that the body becomes lean and thin and 51% respondents reported weakness and lethargy (Table 4.10), showing awareness of the disease and its symptoms. In addition to these symptoms, we also enquired from the respondents how would they know and detect if they have diabetes and 55% and 46% respondents said they would undergo a blood test and urine test, respectively, as the diagnostic test for diabetes (Table 4.10).

Table 4.9 Sources of awareness of diabetes among respondents

Proportion of respondents who had ever heard of diabetes	**Baseline**	**End line**
	38.1	96.3
	Frequency	**Percentage**
Where did you hear about diabetes?		
Number of respondents reported—Radio/TV	33	8
Number of respondents reported—Newspaper, magazine	24	6
Number of respondents reported Doctor	77	19
Number of respondents—Nurse, pharmacist, lab technician	14	3
Number of respondents—Health worker	37	9
Number of respondents—Friends and relatives	239	59
Number of respondents—Project staff	285	70
IEC Activities like street plays, wall writings, pamphlets	178	44
Number of respondents—Camps	53	13
Number of respondents—Others	0	0
Number of respondents—I don't remember	2	0
	Baseline	**End line**
Proportion of respondents aware of diabetes as a disease	24.7	96.4
Proportion of respondents aware of at least TWO symptoms of diabetes	12.7	61.2
	Frequency	**Percentage**
Reaction to diabetes (What happens to the body if one has diabetes)		
Number of respondents who had 'High sugar in blood'	212	52
Number of respondents who had 'High sugar in urine'	92	23
Number of respondents who reported 'Others' (disease symptoms)	104	25
Number of respondents who reported 'I have no idea'	15	4

Risk Factors

During the end line survey, the risk factors were spelled out clearly by the respondents. Among the risk factors, high carbohydrate diet, irregular food intake, alcoholism and pregnancy were considered as the major threat by the respondents.

With reference to Table 4.11, the respondents rated the factors related to the prevention of disease namely abstaining from alcohol (59.3%), limiting carbohydrate diet (45.9%), regular exercise (42.7% and 57.3% for females and males, respectively) and intake of diet in regular intervals in Table 4.12. The end line survey results proved again that KAP helps in the prevention and control of the disease.

Table 4.10 Symptoms awareness of diabetes (end-line)

		Female		*Male*	
		Count	*Row total* N (%)	*Count*	*Row total* N (%)
Disease symptoms					
B1.5a. Excessive appetite	Yes	103	56.9	78	43.1
	No	82	41.4	116	58.6
B1.5b. Excessive thirst	Yes	54	53.5	47	46.5
	No	131	47.1	147	52.9
B1.5c. Excessive urination	Yes	105	51.2	100	48.8
	No	80	46.0	94	54.0
B1.5d. Body becomes lean and thin	Yes	28	41.8	39	58.2
	No	157	50.3	155	49.7
B1.5e. Weakness/lethargy	Yes	79	48.8	83	51.2
	No	106	48.8	111	51.2
B1.5f. I have no idea	Yes	35	58.3	25	41.7
	No	150	47.0	169	53.0
B1.5x. Other	Yes	14	48.3	15	51.7
	No	171	48.9	179	51.1
B1.5x. Other specifics (e.g. *Vertigo, Chest pain, Injury, Blisters, Sore wounds, Swollen legs, etc.*)		191	50.4	188	49.6
Detecting sugar					
B1.6a. By doing a blood test	Yes	122	45.0	149	55.0
	No	63	58.3	45	41.7
B1.6b. By doing urine test	Yes	47	54.0	40	46.0
	No	138	47.3	154	52.7
B1.6c. I have no idea	Yes	43	59.7	29	40.3
	No	142	46.3	165	53.7
B1.6x. Other	Yes	3	42.9	4	57.1
	No	182	48.9	190	51.1

Undergoing Diagnostic Test

The respondents were also measured for the diagnostic test for the disease after the baseline survey, and there was an increase in the percentages of the respondents for the blood sugar test to detect the disease. Table 4.13 shows that the end line increased to 37 from baseline 21.4. Before the commencement of the project, there were only 10 respondents who underwent blood sugar test, but after the intervention the numbers swelled up, in the case of male respondents it was 72 and in the case of female respondents the number increased to 59. Moreover, male

Table 4.11 Awareness of risk factors and prevention of diabetes

Proportion of respondents who knew at least two risk factors of diabetes				**Baseline** 13.1	**End line** 59.9
Awareness of risk factors and prevention of diabetes—End line		**Female**		**Male**	
		Count	**Row total** *N* (%)	**Count**	**Row total** *N* (%)
B1.7a. Overweight	Yes	38	55.9	30	44.1
	No	147	47.3	164	52.7
B1.7b. Age above 45	Yes	39	72.2	15	27.8
	No	146	44.9	179	55.1
B1.7c. High blood pressure	Yes	17	81.0	4	19.0
	No	168	46.9	190	53.1
B1.7d. Lack of exercise	Yes	14	42.4	19	57.6
	No	171	49.4	175	50.6
B1.7e. Stress	Yes	20	54.1	17	45.9
	No	165	48.2	177	51.8
B1.7f. Alcoholism	Yes	24	40.7	35	59.3
	No	161	50.3	159	49.7
B1.7g. Irregular food intake	Yes	93	45.8	110	54.2
	No	92	52.3	84	47.7
B1.7h. High carbohydrate diet	Yes	74	42.5	100	57.5
	No	111	54.1	94	45.9
B1.7i. During pregnancy	Yes	1	33.3	2	66.7
	No	184	48.9	192	51.1
B1.7j. I have no idea	Yes	39	54.2	33	45.8
	No	146	47.6	161	52.4
B1.7x. Other	Yes	2	40.0	3	60.0
	No	183	48.9	191	51.1

respondents preferred to have their investigations from the private clinic, whereas the females preferred to have it from government institutions. The NGO-run clinics and camps were also preferred for the investigations, as they are primarily free of cost. Also, 50 male respondents and 54 female respondents had undergone diagnostic test more than two times.

Prevention of the Disease

The international diabetes forum recommends six ways for the prevention of the disease.

The interventions suggested/recommended were taken seriously as can be seen from the following results. All the 19 respondents who underwent diagnostic procedures followed their doctor's advice sincerely and worked

Table 4.12 Protecting self from diabetes end line

Prevention and control of disease		*Female*		*Male*	
		Count	*Row total* N *(%)*	*Count*	*Row total* N *(%)*
B1.8a. Reduction of weight and	Yes	35	56.5	27	43.5
preventing weight gain	No	150	47.3	167	52.7
B1.8b. Control of high blood	Yes	5	55.6	4	44.4
Pressure	No	180	48.6	190	51.4
B1.8c. Regular exercise	Yes	35	42.7	47	57.3
	No	150	50.5	147	49.5
B1.8d. Reduction of stress	Yes	17	54.8	14	45.2
	No	168	48.3	180	51.7
B1.8e. Intake of diet in regular	Yes	114	49.1	118	50.9
interval	No	71	48.3	76	51.7
B1.8f. Limiting/stopping alcohol	Yes	22	41.5	31	58.5
intake	No	163	50.0	163	50.0
B1.8g. Stop taking too much of	Yes	81	48.2	87	51.8
carbohydrate in diet	No	104	49.3	107	50.7
B1.8h. Regular checking of blood	Yes	12	63.2	7	36.8
sugar after the age of 45	No	173	48.1	187	51.9
B1.8i. Blood sugar test during	Yes	1	100.0	0	0.0
pregnancy	No	184	48.7	194	51.3
B1.8j. I have no idea	Yes	31	58.5	22	41.5
	No	154	47.2	172	52.8
B1.8x. Other	Yes	8	50.0	8	50.0
	No	177	48.8	186	51.2

on them. On asking the question, whether you follow the doctor's advice, majority (3.1 females and only 26.9 males) of the respondents agreed to it.

From Table 4.14, it is clear that 14.5% female and 12.7% male respondents replied that after the age of 40 years they feel that blood sugar check-up is necessary.

Advise to Friend/Relative

From Table 4.15, it is clear that nearly 31 per cent female respondents and 32 per cent male respondents feel they will recommend others to reduce weight and consult doctors regularly for the same. Also, for the prevention of diabetes, 195 females and 18% males say that they will advise friends/relatives to control sugar and 39% female and 45% male respondents recommend that they advocate friends/relatives to consult doctors regularly for diabetes.

Table 4.13 Testing of blood sugar test

		Female		Male	
		Count	*Row* N (%)	*Count*	*Row* N (%)
B1.9. Have you ever gone through blood sugar test	Yes	69	45.7	82	54.3
	No	136	52.9	121	47.1
B1.10. When did you get blood tested for sugar	Before the project 2012	10	50.0	10	50.0
	After the project 2013	59	45.0	72	55.0
Where did you get the blood tested for sugar?					
B1.11a. Get Blood sugar tested: Govt. hospital/ testing centre	Yes	24	44.4	30	55.6
	No	45	46.4	52	53.6
B1.11b. Private hospital/ testing centre	Yes	16	40.0	24	60.0
	No	53	47.7	58	52.3
B1.11c. NGO-run clinic	Yes	28	54.9	23	45.1
	No	41	41.0	59	59.0
B1.11d. Camp	Yes	27	47.4	30	52.6
	No	42	44.7	52	55.3
B1.11e. Mobile clinic	Yes	0	0.0	3	100.0
	No	69	46.6	79	53.4
B1.11x. Other	Yes	4	50.0	4	50.0
	No	65		78	54.5
B1.12. How many times have you gone through the blood sugar test					
	Only 1 time	21	40.4	31	59.6
	2 times	17	39.5	26	60.5
	3 times	12	41.4	17	58.6
	>3 times	15	68.2	7	31.8
	Don't remember how many times	4	80.0	1	20.0

7 Conclusion and Recommendations

Since India has been declared as the diabetic capital of the world, there is an enormous economic, social and personal cost of type 2 diabetes making the disease an important case for prevention. In recent years, there has been much new evidence demonstrating the potentially preventable nature of type 2 diabetes, particularly by the implementation of lifestyle measures such as weight control and exercise (Steyn et al. 2004) apart from knowledge attitude and practice towards the disease. Since this is a baseline and end line survey hence the conclusions and recommendations of the study, the author feels and advocates strongly, to be used as the policy guideline for prevention of T2DM.

Table 4.14 Blood sugar detection

	Female		*Male*	
	Count	*Row total* N (%)	*Count*	*Row total* N (%)
B1.13. Has your doctor/project staff ever told you after the blood test?				
Yes	19	67.9	9	32.1
No	50	40.7	73	59.3
B1.14. Are you following what your doctor/project staff has advised you to do?				
Yes	19	73.1	7	26.9
No	0	0.0	2	100.0
B1.18. Has the project staff ever asked you to develop a kitchen garden?				
Yes	127	51.0	122	49.0
No	78	49.1	81	50.9
B1.19. Do you have kitchen garden at home?				
Yes	43	47.8	47	52.2
No	162	50.9	156	49.1
B1.20 After what age (years) do you think one should go for sugar check-up regularly?				
After 30	32	45.1	39	54.9
After 40	59	53.2	52	46.8
After 50	17	40.5	25	59.5
After 60	3	75.0	1	25.0
At any age	43	55.1	35	44.9
Not required if person is fit	0	0.0	1	100.0
No idea	51	52.0	47	48.0
Other	0	0.0	3	100.0
Advise follow up				
B1.15. Why are you not following your doctor's/project staff's advice?				
I have no physical complaints	0	0.0	2	100.0
I didn't go for follow-up visit to doctor after blood test	0	0.0	0	0.0
Family refusal	0	0.0	0	0.0
No camps	0	0.0	0	0.0
B1.20 After what age (years) do you think one should go for sugar check-up regularly?				
After 30	32	7.8	39	9.6
After 40	59	14.5	52	12.7
After 50	17	4.2	25	6.1
After 60	3	0.7	1	0.2
At any age	43	10.5	35	8.6
Not required if the person is fit	0	0.0	1	0.2
I have no idea	51	12.5	47	11.5
Other	0	0.0	3	0.7

Table 4.15 Willingness to give advice and help

Advice					**End-line**
Proportion of respondents who are willing to advice their overweight relative/family member/colleague 'to reduce weight'					66.9
Proportion of respondents who are willing to advice their diabetic relative/family member/colleague 'to control sugar'					38.0
		Female		**Male**	
		Count	**Table *N* (%)**	**Count**	**Table *N* (%)**
Advice to friend/relative/family member—Overweight					
B1.21a. Why should I give him/her advice	Yes	1	0.2	2	0.5
	No	204	50.0	201	49.3
B1.21b. Advise him/her to reduce weight	Yes	128	31.4	145	35.5
	No	77	18.9	58	14.2
B1.21c. Advise him/her to consult the doctor regularly	Yes	126	30.9	132	32.4
	No	79	19.4	71	17.4
B1.21d. I have no idea what to advise	Yes	32	7.8	7	1.7
	No	173	42.4	196	48.0
B1.21e. Advise to develop kitchen garden	Yes	8	2.0	13	3.2
	No	197	48.3	190	46.6
Advice to friend/relative/family member—Diabetes					
B1.22a. Why should I give him/her advice	Yes	2	0.5	2	0.5
	No	203	49.8	201	49.3
B1.22b. Advise him/her to control his/her sugar	Yes	81	19.9	74	18.1
	No	124	30.4	129	31.6
B1.22c. Advise him/her to consult the doctor regularly	Yes	159	39.0	184	45.1
	No	46	11.3	19	4.7
B1.22d. I have no idea what to advise	Yes	24	5.9	6	1.5
	No	181	44.4	197	48.3
B1.22e. Advise to develop kitchen garden	Yes	15	3.7	17	4.2
	No	190	46.6	186	45.6

Conclusion

Diabetes mellitus is reaching epidemic proportions in India. The level of morbidity and mortality due to diabetes and its potential complications are enormous and pose significant healthcare burdens on both families and society. A study, however, shows a slight decline in deaths from communicable diseases such as malaria and tuberculosis, while chronic diseases such as cancer and diabetes are on the rise (*The Times of India* June 2015).[1]

[1]According to the World Health Organization (WHO), India had 69.2 million people living with diabetes in 2015.

Nearly 87% of the people went for a blood test after the intervention under the project. The major contribution by the intervention study can be seen by the fact that a large number of people who had blood sugar more than the normal value (hyperglycaemia) had been able to control their blood sugar levels with exercise and dietary measures as promoted under the project and many of the mildly affected had been able to control it to such an extent, that they did not need anti-diabetic oral medicines.

Government policies help in creating guidelines on diabetes management, funding community programmes for public awareness about diabetes risk reduction and availability of medicines and diagnostic services to all sections of the community. An effort by various governments and agencies around the world to intervene in diabetes management has resulted in positive health outcomes for their communities.

Recommendations

The following recommendations are suggested for awareness creation and prevention intervention projects for diabetes.

1. Regular and constant monitoring could play an important role in the follow-up of the patients suffering from diabetes. The programme should emphasise on regular screening of people with risk factors and should advise them to go for a screening test on empty stomach.
2. Based on screening criteria, the suspect cases should be referred to the nearest Community Health Centre (CHC) for investigation and confirmation of diabetes.
3. More aggressive Behavioral Change Communication (BCC) should be planned to bring about behavioural changes. The IEC material developed by the organisation under the project should be displayed in sub-centres and PHCs and at prominent places. This requires regular monitoring by the staff residing in the respective villages.
4. NCD cell at district/CHC has a very important role in the prevention and control of diabetes. Therefore, these cells should be strengthened in terms of HR and facilities, and made functional.
5. The problem of attitude towards the disease and health-seeking behaviour among women need focused BCC and this should be made an integral part of BCC campaign.
6. The findings of the evaluation of the project are encouraging in terms of an increase in knowledge and influencing the attitudes and

behaviour of the population in dealing with diabetes. The intervention deserves to be continued for institutionalisation and is a good practice for up-scaling.

7. As discussed with various health officials, the main concern was that the NCD cell was weak and needed rejuvenation. The efforts of Humana People to People India in the whole district can contribute a lot and not only in the area of diabetes but also other diseases like cancer, malaria, HIV/AIDS and in the anti-tobacco campaign.

References

Anjana, R.M., Pradeepa, R., Deepa, M. et al. (2011). Prevalence of Diabetes and Prediabetes (Impaired Fasting Glucose and/or Impaired Glucose Tolerance) in Urban and Rural India: Phase I Results of the Indian Council of Medical Research–INdia DIABetes (ICMR–INDIAB) study. *Diabetologia, 54*, 3022–3027. https://doi.org/10.1007/s00125-011-2291-5.

Darnton-Hill, I., Nishida, C., & James, W. P. (2004). A Life Course Approach to Diet, Nutrition and the Prevention of Chronic Diseases. *Public Health Nutrition, 7*(1A), 101–121.

Dobe, M. (2012). Health Promotion for the Prevention and Control of Non Communicable Diseases: Unfinished Agenda. *Indian Journal of Public Health, 56*(3), 180–186.

Dowse, G. K., Gareeboo, H., Alberti, K. G., Zimmet, P., Tuomilehto, J., Purran, A., Fareed, D., Chitson, P., & Collins, V. R. (1995). Changes in Population Cholesterol Concentrations and Other Cardiovascular Risk Factor Levels after Five Years of the Noncommunicable Disease Intervention Programme in Mauritius. *British Medical Journal, 311*(7015), 1255–1259.

Fuentes, R., Fuentes, R., Ilmaniemi, N., Laurikainen, E., Tuomilehto, J., & Nissinen, A. (2000). Hypertension in Developing Economies: A Review of Population-based Studies Carried out from 1980 to 1998. *Journal of Hypertension, 18*(5), 521–529.

Gupta, R., & Gupta, K. D. (2009). Coronary Heart Disease in Low Socioeconomic Status Subjects in India: "An Evolving Epidemic". *Indian Heart Journal, 61*(4), 358–367.

Hetzel, B., & McMichael, A. (1987). *The Lifestyle Factor: Lifestyle and Health*. Melbourne: Penguin.

ICMR-India DIABetes[INDIAB] study, Phase 1. (2011). *India Diabetes (ICMR-INDIAB) Study of 2011*. New Delhi: Indian Council of Medical Research.

India's diabetes rate up 123% since 1990. (2015, June). *The Times of India*. Retrieved from https://timesofindia.indiatimes.com/india/Indias-diabetes-rate-up-123-since-1990/articleshow/47670045.cms.

Joshi, S. R., Das, A. K., & Vijay, V. J. (2008). Challenges in Diabetes Care in India: Sheer Numbers, Lack of Awareness and Inadequate Control. *The Journal of Association of Physicians India, 56*, 443–450.

Katan, M. B., Zock, P. L., & Mensink, R. P. (1995). Dietary Oils, Serum Lipoproteins, and Coronary Heart Disease. *The American Journal of Clinical Nutrition, 61*(6 Suppl), 1368S–1373S.

King, H., Aubert, R. E., & Herman, W. H. (1998). Global Burden of Diabetes, 1995–2025: Prevalence, Numerical Estimates, and Projections. *Diabetes Care, 21*, 1414–1431.

Matsudo, S. M., & Matsudo, V. R. (2004). Physical Activity Promotion: Experiences and Evaluation of the Agita Sao Paolo Program using the Ecological Mobile Model. *Journal of Physical Activity and Health, 1*, 81–97.

McQueen, D. V. (2013). *Global Handbook on Noncommunicable diseases and Health Promotion*. New York: Springer. https://doi.org/10.1007/978-1-4614-7594-1.

Mendis, S. (2010). The Policy Agenda for Prevention and Control of Non Communicable Diseases. *British Medical Bulletin, 96*, 23–43.

Murugesan, N., Snehalata, C., Shobana, R., Roglic, G., & Ramachandran, A. (2007). Awareness About Diabetes and its Complications in the General and Diabetic Population in a City in Southern India. *Diabetes Research and Clinical Practice, 77*, 433–437. https://doi.org/10.1016/j.diabres.2007.01.00.

Narula, D. (2016). Factors Influencing T2DM—A Case Study of Jaipur, District, Rajasthan. *IJMER, 5*(9), 190–217.

Nissinen, A., Berrios, X., & Puska, P. (2001). Community-based Noncommunicable Disease Interventions: Lessons from Developed Countries for Developing Ones. *Bulletin of the World Health Organization, 79*(10), 963–970.

Penny, M., & Etherton, K. (1999). Monosaturated Fatty Acids and Risk of Cardiovascular Disease Circulation. *PubMed, 100*(11), 1253–1258.

Prabhakaran, D., Jeemon, P., & Goenka, S. (2009). Impact of a Worksite Intervention Program on Cardiovascular Risk Factors: A Demonstration Project in an Indian Industrial Population. *Journal of American College of Cardilogy, 53*, 1718–1728. https://doi.org/10.1016/j.jacc.2008.12.062.

Purran, A., Dowse, G. K., Hassam, G., Alberti, M. M., George, K. (1995). Changes in Population Cholesterol Concentrations and other Cardiovascular Risk Factor Levels After Five Years of the Non-communicable Disease Intervention Programme in Mauritius. *BMJ, 311*, 1255. https://doi.org/10.1136/bmj.311.7015.1255.

Saraf, D. N. (2012). A Systematic Review of School-based Interventions to Prevent Risk Factors Associated with Noncommunicable Diseases. *Asia-Pacific Journal of Public Health, 24*(5), 733–752.

Seema, K. A., & Cornwall, J. (2014). The Current State of Diabetes Mellitus in India. *Australasian Medical Journal, 7*(1), 45–48. https://doi.org/10.4066/AMJ.2013.1979.

Singh, K., & Reddy, K. S. (2011). What are the Evidence Based Public Health Interventions for Prevention and Control of NCDs in Relation to India? *Indian Journal of Community Medicine, 36*(5), 23–31.

Somannavar, S., Lanthorn, H., & Deepa, M. (2008). Increased Awareness about Diabetes and its Complications in a Whole City: Effectiveness of the 'Prevention, Awareness, Counselling and Evaluation' [PACE] Diabetes Project [PACE-6]. *The Journal of Association of Physicians of India, 56*, 497–502.

Steyn, N. P., Mann, J., & Bennett, P. H. (2004, February). Diet, Nutrition and the Prevention of Type 2 Diabetes. *Public Health Nutrition, 7*(1A), 147–165.

Traversy, G., & Jean-Philippe, C. (2015). Alcohol Consumption and Obesity: An Update. *Current Obesity Reports (Springer), 4*(1), 122–130.

U.S. Department of Health and Human Services. (2010). *How Tobacco Smoke Causes Disease: What It Means to You*. Atlanta: Centers for Disease Control and Prevention, National Center for Chronic Disease Prevention and Health Promotion, Office on Smoking and Health. Retrieved May 5, 2014, from https://dcp.psc.gov/ccbulletin/articles/tobacco_01_2011.aspx, https://das.nh.gov/wellness/docs/HowTobaccoSmokeCausesDisease.pdf

Wilson, C. (2010). Eating, Eating is Always There: Food, Consumerism and Cardiovascular Disease. Some Evidence from Kerala, South India. *Anthropology & Medicine(NCBI), 17*, 261–275. https://doi.org/10.1080/13648470.2010.526699.

World Health Organization. (2004). *Global Strategy on Diet, Physical Activity and Health*. Geneva: World Health Organization.

World Health Organization. (2014). *Global Status Report on Noncommunicable Diseases 2014*. Selangor, Malaysia: WHO.

Yadav, K., & Krishnan, A. (2008). Changing Patterns of Diet, Physical Activity and Obesity among Urban, Rural and Slum Populations in North India. *Obesity Reviews, 9*(5), 400–408. https://doi.org/10.1111/j.1467-789X.2008.00505.x.

CHAPTER 5

Assessing the Impact of Government Schemes on Rural Development: A Study of Udhampur District (Jammu and Kashmir State)

Deepti Abrol, Jhanvi Khanna, Kamani Dutta, and Ritika Sambyal

1 Introduction

In India people predominantly live in rural areas, and the rural economy continues to be the major source of employment for a majority of the people. Furthermore, employment has linkage with poverty, especially in rural areas. Besides this, India is an agrarian economy where more than 50

D. Abrol (✉)
Department of Commerce, P.G. College for Women,
Gandhi Nagar, Jammu, Jammu and Kashmir, India

J. Khanna
Jammu and Kashmir State Board of School Education,
Jammu, Jammu and Kashmir, India

K. Dutta • R. Sambyal
Department of Commerce, Udhampur Campus, University of Jammu,
Jammu, Jammu and Kashmir, India

H. Chahal et al. (eds.), *Sustainable Business Practices for Rural Development*, https://doi.org/10.1007/978-981-13-9298-6_5

per cent of its working population is dependent on agriculture for their livelihood. Even after unprecedented levels of migration of people from rural to urban areas due to various reasons, a majority of the Indian population, that is, around 70 per cent of the population, continues to live in rural areas, which means out of 121 crore Indians, 83.3 crore live in the rural areas (Census 2011). The Central and the State Government have taken initiatives to develop and implement various schemes in India for the welfare of their citizens from time to time. The 73rd Constitutional Amendment was extremely imperative for the rural population of Indian villages, as most of the schemes were developed and implemented for the overall advancement of rural areas and people (Naskar and Kumar 2017). Many decades ago Mahatma Gandhi stated that India resides in its villages and, ironically, even after seven decades of independence, the condition remains the same. Even today when someone talks about India, it is said that if you want to see actual India, you must visit its villages. The rural India shown in movies is beautiful, relaxing and picturesque. But those images depict only half of the picture and do not show the challenges and problems faced by people living in those areas. Rural population in India still lives in primitive conditions and has regressed in almost every socio-economic aspect. They are facing many difficulties as far as education, health, employment, social welfare and so on are concerned. Around three-fourths of the rural families earn a meagre amount of monthly income which is less than Rs 5000. Also, more than half the rural households do not have their own land and are casual labourers (Socio-Economic Census 2011). Over the years, though the level of poverty in our country has reduced, the rate of poverty reduction in rural areas has remained quite less as compared to urban areas. According to the Rangarajan Committee, in 2011–2012, rural poverty was more than urban poverty and stood at 31 per cent approximately.

A healthy mind, body and soul are pre-requisites for the efficient human capital and can be acquired only through good health and adequate nutrition. But the present situation is such that rural areas tend to lag and thus a huge gap is created between the prosperous urban areas and the decaying poor rural areas. Therefore, to overcome all these problems, to improve their conditions and to bring rural population at par in the society, the government is making continuous efforts and formulating various schemes at both Central and State levels. To satisfy the needs of the rural poor, government is laying emphasis on poverty alleviation, generation of employment, provision of basic amenities and social welfare. The primary objective of the ministry is to promote and assist voluntary actions for implementing government schemes for attaining rural prosperity.

It has been seen that rural development can be attained only when government's efforts are adequately supplemented by both direct and indirect involvement of people at the ground level. For achieving the objective of developing rural areas, the Government of India has employed various schemes such as Indira Awas Yojana (IAY), MGNREGA, National Rural Livelihoods Mission (NRLM), Pradhan Mantri Gram Sadak Yojana (PMGSY), Saansad Adarsh Gram Yojana (SAGY) and many more. Though there are several studies conducted on the various government schemes from time to time (specifically on MGNREGA), most of them have a thrust on the role of individual schemes. For example, Naskar and Kumar (2017) examined how far the Pradhan Mantri Jan Dhan Yojana (PMJDY) is accepted by the rural people in the rural tribal populated area. The study highlighted PMJDY has played an important role in improving financial inclusion, in which every individual can get access to credit, insurance and pension facility through the banking system. The authors of this study further mentioned that there is a need to create awareness about the benefits of PMJDY among each individual of the community for the economic development of rural poor. Bhargava (2014) analysed the effects of National Rural Employment Guarantee Act (NREGA) on rural labour markets after technology adoption decisions taken by farm owners. Specifically, the study examined the impact of fertiliser subsidies on input utilization, agricultural output and other household outcomes. The results revealed the substantial and persistent impact on all these outcomes. Sami and Khan (2016) ponder upon the need to see the impact and efficiency of different government schemes on rural development. They further remarked on better management and modifications in the government schemes to make them more fruitful. Dodd et al. (2017) opined that the need of the hour is to create more awareness in the villages as far as different government schemes are concerned. Manoj and Alexon Puthukkeril (2015) stressed upon the problems and prospects of MGNREGA with respect to women labourers and suggested suitable remedial strategies for more effective implementation of MGNREGA. Jain (2017) studied the socio-economic impact of MGNREGA on rural workers and provided essential evidence that the programme has produced imperative short- and long-run role of this scheme on labour well-being, wages and so on. Besides this, Dreze (2005) contends that MGNREGA can produce an impact on the state and give haggling energy to the workers. Further, in this regard, Bhatty (2006) and Krishnamurthy (2006) suggested focusing on MGNREGA and other related schemes to provide solutions for emergencies, especially the crisis that happens locally. Karthika (2015) discussed the benefits of MGNREGA and focused on the impact of this programme on rural development and women empowerment and demonstrated a significant effect of this scheme. De et al. (2016) focused on sanitation coverage in the rural areas for the better rural health of the people living in rural areas. The results publicised that the scheme of Swachh Bharat Mission enhances social skills such as communication, mobility, participation, decision making and so on. Recently, Shaik (2017) explained how utilisation of scarce resources effectively influences the development of rural villages. After reviewing the existing literature, it has been realised that maximum research was conducted on individual government schemes such as MGNREGA, PMGSY, Indira Awaas Yojana and so on. Therefore, the present study shall fill the existing void in literature and endeavours to highlight the impact of three major government schemes (e.g., MGNREGA, Swachh Bharat Mission and Pradhan Mantri Awas Yojana) that focus on rural development measured through economic and social development.

Specifically, the research question that shall be addressed in the present study is: What is the role of various government schemes in the development of rural areas?

2 Conceptual Framework

From time to time Indian Government has been announcing various government schemes at all levels for a large section of the society. These schemes could be either Central, state-specific or collaborations between the Centre and the states. These schemes are implemented for both economic and social welfare of the society. These schemes aim to reduce poverty, unemployment and income inequalities as far as economic welfare is concerned. It also focuses on increasing social welfare of the people by providing them education, health and sanitation facilities and special schemes for underprivileged sections of the society and bringing them on par with other members of the society by reducing regional disparities. Since independence, the Government of India has acknowledged the need to develop villages of rural areas for economic and social growth of the country and its citizens. In the present study, selected villages of Udhampur district were considered in which only three government schemes were being implemented namely MGNREGA, Pradhan Mantri Awas Yojana-Gramin (PMAY-G) and Swachh Bharat Mission. The government schemes undertaken in the study are discussed in brief in the following sections.

MGNREGA

In the year 2005, the Government of India initially introduced a scheme of NREGA which is now more commonly known as MGNREGA. Passed as an Indian Labour Law on 2 February 2006, MGNREGA was initially implemented in 200 districts across India, and later on 1 April 2008 some more districts were added to the earlier list. It is a social safety measure with an objective to provide livelihood, sustenance, social protection, empowerment to the socially underprivileged and employment to the people living in rural areas with an aim to provide assurance of right to work as a source of revenue. Apart from ensuring economic security and generating assets for rural people, NREGA scheme can assist in protecting the surroundings, empowering the women living in rural areas, sinking migration of people from rural areas to urban areas and developing social equity among people of rural areas. This scheme is applicable for all rural adults who have volunteered for unskilled manual labour and can get a minimum 100-day job guarantee in every financial year. According to the World Development Report of 2014, NREGA has been implemented on a large scale and is an excellent example of rural development.

Pradhan Mantri Awaas Yojana-Gramin (PMAY-G)

In January 1996, as a tool of poverty alleviation and rural housing programme, IAY was started. Although IAY focused on the housing needs of rural households, there were certain gaps such as non-assessment of housing, lack of transparency in selection of beneficiaries, distribution of loans, poor quality of houses, lack of administration and supervision which restricted the impact and results of the programme. To address these gaps and in view of the government's aim of providing 'Housing to All' by 2022, Indira Awaas Yojana was reframed as Pradhan Mantri Awaas Yojana-Gramin (PMAY-G) on 1 April 2016. For taking the benefit of this scheme, the applicant's age must be 70 years or less, and the income should be below Rs 3 lakhs per annum for economically weaker sections and between Rs 3 and 6 lakhs per annum for lower income groups. The applicant's family should comprise a husband, wife and unmarried children; however, under Middle Income Group (MIG) category, an adult earning member irrespective of his/her marital status can be treated as a separate household. In case of married couples, either of the spouse or both of them together in joint ownership can become eligible for single subsidy. The beneficiaries of this scheme are entitled for interest subsidy on the purchase or construction of a house. By the year 2022, the scheme aims to provide pucca house with all basic facilities to all houseless people. Moreover, the beneficiaries can also get Rs 90–95 per day of unskilled labour from MGNREGA.

Swachh Bharat Abhiyan

The honourable prime minister of India, Shri Narendra Modi, initiated the Swachh Bharat Mission on 2 October 2014 for achieving universal sanitation coverage and cleanliness drive throughout the country. This scheme aims to achieve a target of clean India by 2019 as a tribute to Mahatma Gandhi on his 150th birth anniversary. This implies improving the level of cleanliness in the country through solid and liquid waste management activities and also to make the Indian villages Open Defecation Free (ODF). It is a step towards improving the overall quality of life of people living in India. It motivates the people to use cost effective technologies for ecologically safe and sustainable sanitation through health education and mass awareness programmes.

In the present chapter, the government schemes undertaken aim to have an impact on the social and economic development of the rural areas. Therefore, we shall discuss the theoretical background of the outcomes.

Rural Development

Rural development can be defined as the overall development of the rural areas which is based on effective and optimum utilisation of India's natural resources through scientific measures. Several efforts have been made for developing the rural areas for better social and economic development from the time of ancient history to the time of British rule (Shaik 2017). Thus, rural development means desired positive change in the rural areas—both in a quantitative and qualitative sense. In India, rural development is one of the most important factors for the growth of Indian economy. The main aim of developing rural areas is to improve the standard of living of rural people by providing them with food, shelter, education, health services, education, employment and empowerment. Moreover, another significant objective of the scheme is to enhance productivity in rural areas and reduce poverty. Thus, rural development means both economic and social betterment of the people.

Economic Development

Economic development can be defined as the development of economic wealth of the country for the well-being of their citizens. It aims to improve the economic condition and quality of life of the people by creating various employment opportunities. Broadly, it focuses on major areas such as government policies, price stability, high employment, infrastructural facilities and services like housing, highways, educational programmes and so on. It aims at providing employment opportunities for the poor and low-income households for economic growth.

Social Development

Social development has been defined as a process of transformation from traditional way of living of rural communities to progressive ways of living. It is a way through which people can be assisted to develop their capacities and resources on their own. It is directly concerned with the investment in human capital and aims at total development of the people by providing social justice. Social development has been defined by scholars as the promotion of a sustainable society that is worthy of human dignity by empowering marginalised groups, men and women, to undertake their own development, to improve their social and economic position and to acquire their rightful place in the society.

3 Research Objective

The main objective of this research is to measure and assess the direct impact of government schemes on rural development which includes both economic and social development in rural areas of Udhampur district (Jammu and Kashmir). The government schemes which are implemented in Udhampur district and considered for the study are MGNREGA, PMAY-G and Swachh Bharat Mission.

4 Hypotheses Development

The development in the rural areas is an indispensable part of the country's social and economic development. The objective of rural development is the fortification of the superiority of human life in rural and backward areas accompanied by bridging the gap between the urban and rural areas by providing all amenities to their people. Various government initiatives, policies and programmes for the successful development of rural areas should aim at diversification of economic and non-economic activities by minimising the dependence of people of these areas only on agriculture and bringing about a necessary augmentation in the share of associated activities, industries in rural areas, business and service mechanism of the rural economy (Patel et al. 2018). Further, rural development has been judged as the utmost priority in the country's planning process. The PMJDY scheme is completely supportive for the people residing in rural and urban areas for attaining financial services directly from government officials (Shettar 2016). Moreover, Karthika (2015) found that MGNREGA has played a significant role and has brought a positive impact on the rural development and has a potential to enhance the local development (Poonia 2012). Furthermore, Sugapriyan and Prakasam (2015), highlighted that MNREGA is a step to overcome poverty by assuring employment in rural areas to all those who need work and are capable in formation of capital. Pamecha and Sharma (2015) showed that the MGNREGA programme has brought a revolutionary change in the lives of the people living in rural areas. Das (2016) found that MGNREGA has a significant role in inclusive growth of rural areas.

Further, Mathur (2007) concluded that MGNREGA scheme has remarkably initiated the vital transformations in the lives of poor people of rural areas and revealed that migration has decreased in various villages of states such as Andhra Pradesh, Chhattisgarh and Odisha. Shah (2007) estimated an increase of 6.5 per cent in agricultural wages and also found a one standard deviation increase in infrastructure development due to MNREGA and led to a 30 per cent fall in wage sensitivity to production shocks. Tiwari and Upadhyay (2012) in their study titled, 'Impact of NREGA on empowerment of the beneficiaries in West Bengal', found a positive impact on the empowerment of the respondents in the state of West Bengal of MNREGA. Das (2012) highlighted the impact of MGNREGA on women beneficiaries. Kumar and Bhattacharya (2013) in their study titled, 'Participation of women in MGNREGA: How far is it successful in Morigoan, Assam', studied the level of women participation in MGNREGA activities in comparison to males. Again, Arora et al. (2013) in their study titled, 'Mahatma Gandhi National Rural Employment Guarantee Scheme: A Unique for Indian Rural Women', highlighted the importance of implementing MGNREGA scheme on women empowerment residing in Rohtak district of Haryana. Besides this, Farooqi and

Saleem (2015) in their research paper titled, 'Mahatma Gandhi National Rural Employment Guarantee Act (MGNREGA) and Empowerment of Women from BPL families in rural areas: A case study of district Aligarh (India)', found that a provision should be made for sparing women from hard manual work and to extend MGNREGA programme to skilled or semi-skilled work so that women become professionally skilled. Shah (2007) has claimed that NREGA has got the requisite potential to give a 'big push' to the distressed regions in India. It has been pointed out that the multiplier effects of MGNREGA (then NREGA) were significantly and positively contributing to the overall and long-term agricultural development which ultimately reduces the poverty in our country.

Therefore, based on the previous arguments, we have hypothesised that:

Hypothesis 1 : Government schemes significantly affect economic development.

Hypothesis 2 : Government schemes significantly affect social development.

Hypothesis 3 : Government schemes significantly affect overall rural development.

5 Research Methodology

Measures

The present research work focused on the Garnai and Malhar village of Udhampur district of Jammu and Kashmir. Self-developed schedule was framed to collect the primary data from the beneficiaries of government schemes and is based on first-hand information. All statements in the schedule were framed on a five-point Likert scale where '1' stands for strongly disagree, '2' stands for disagree, '3' stands disagree nor agree, '4' stands agree and '5' stands for strongly agree. The questionnaire consisted of two sections: the first section includes items related to the demographic profile of the respondents which includes gender, age, qualification, monthly income; the second section was sub-divided under three important heads, that is, government schemes, economic development and social development. The three constructs of the study are framed using self-generated items. There are 25 items under government schemes head which were developed using the research studies of Sami and Khan (2016), Datta and Singh (2012), Krishnamurthy (2006) and Ambasta (2008).

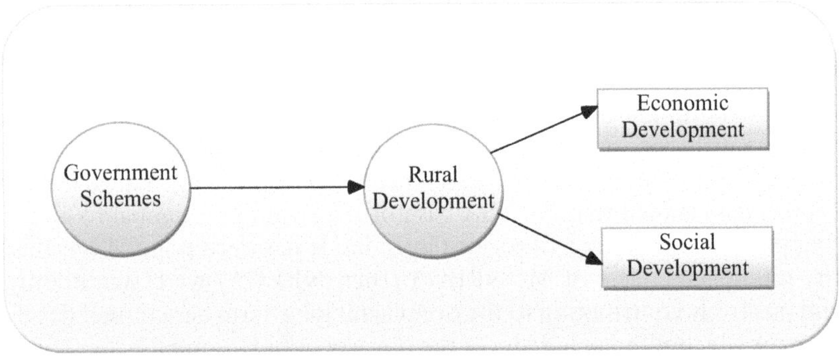

Fig. 5.1 Proposed research model. (Source: Authors)

Economic development scale items (16) were developed from the studies of Pamecha and Sharma (2015) and Pandey et al. (2016). Finally, there are 17 items in innovation scale which were framed from the research studies of Jain (2017) and Bhargava (2014). Moreover, secondary data were collected from the sources like books, newspapers and relevant journals (Fig. 5.1).

Sample and Data Collection

Data for the present study is collected from the beneficiaries of government schemes from Garnai and Malhar village of Udhampur district of Jammu and Kashmir. A field survey using self-developed schedule was conducted. During pilot testing, the study randomly selected 50 respondents from Garnai and Malhar villages (25 each) for collecting information regarding role of government schemes in rural development. The respondents were selected on convenience basis, and the research considered only real beneficiaries of government schemes. On the basis of pretesting results, the final sample size of beneficiaries to be contacted during the final survey was calculated using the statistical formula given by Burns and Bush (2007):

$$n = \frac{S^2 Z^2}{e^2}$$

where '*n*' denotes the sample size; '*Z*' denotes the standard error along with the selected confidence level, that is, 1.96; '*S*' is the variability pointed out by calculated standard deviation; and '*e*' denotes the degree of precision in the estimated sample.

We assumed five per cent margin of error. Therefore, the sample size came to 138 and was rounded off to 140. During the stage of final data collection, respondents were contacted on convenience basis.

Sample Characteristics

Out of the total respondents, 72 per cent beneficiaries were men and 28 per cent were women. The sample of the study has majority of the beneficiaries (58 per cent) that fall in the age group of 20 to 30 years followed by 23 per cent beneficiaries that fall between the age group of 30 and 40 years and the remaining beneficiaries were above 40 years. The level of education was categorised into three levels, that is, up to matric, up to higher secondary and others and the percentage of beneficiaries of government schemes contacted at each level is 71, 29 and 0, respectively. The monthly income of the beneficiaries was divided into four levels. Around 48 per cent of respondents have a monthly income less than Rs 2500, 45 per cent of beneficiaries have income between Rs 2500 and Rs 5000 per month, and the remaining 7 per cent have an income of above Rs 5000 per month.

Normality

After the final survey, the collected data were refined and analysed to derive the significant results by adopting appropriate statistical tools and techniques. Before proceeding for data purification, negative responses assigned to some scale items were reversed. Subsequently, outlier responses of the data were removed, and normality of data was also checked using box-plot method to determine item-to-item outlier and later by using range method. Following this, 16 outliers were determined which were removed from the responses. Finally, the skewness and kurtosis techniques were used to confirm the statistical normalcy of data, whereas Q-Q plot and box plot were used to check the graphical normalcy. All the skewness and kurtosis values were found to be within the threshold criterion (threshold value between +3 and −3), confirming normal distribution of data.

6 Analysis and Results

Exploratory Factor Analysis (EFA)

The multivariate data reduction technique of factor analysis was performed on all constructs, that is, government schemes, social development and economic development. It was performed with Principal Component Analysis method along with orthogonal rotation procedure varimax for summarising the original information into manageable information and minimum factors. Firstly, value of KMO (Kaiser-Meyer-Olkin) above 0.70 and significant BTS value (Bartlett's test of Sphericity) was considered as an indicator of appropriateness of EFA (Malhotra and Dash 2010). The items of the constructs having factor loading value less than 0.5 and Eigen value which is less than 1.0 were ignored for further analysis. In addition, the value of VE (variance explained) above 50 per cent was also taken for further analysis. The results of the EFA are discussed in Table 5.1.

Government Schemes

The process of EFA was performed on government schemes construct which consisted of 25 items. It took around ten rounds to complete. In the first round, 6 factors emerged having 53.324 per cent variance explained. In this round one item is deleted having low communality, that is, below 0.50. The second round again emerged with 6 factors with 56.220 per cent variance explained, and, in this round, again one item with low communality is deleted. Similarly, seven more rounds were completed. In these rounds, items were also deleted on the same criteria. In the final round, the KMO value was accorded at 0.677, and 3 factors, along with identified items emerged. The percentage of variance elucidated by this construct is 62.803 and its Cronbach's alpha value came to be 0.892. The factors included in the construct are accessibility, availability and usage.

Economic Development

The measure 'Economic Development' comprises 16 items and the process got completed in three rounds with deletion of only 3 items having communality value below 0.50 and 3 factors with 4 items in both the first and second factors and finally 5 items in the third factor. The three factors

Table 5.1 Results of exploratory factor analysis (EFA)

Construct	*Factors*	*Items*	*Factor loading value*	*Communality value*	*KMO value*
Government schemes	*Accessibility*	Easy access to the information related to the schemes	0.673	0.523	0.877
		Easy access to the benefits related to the schemes	0.658	0.548	
		Government officials are cooperative, friendly and knowledgeable	0.614	0.518	
		Employees possess sufficient information related to the schemes	0.685	0.540	
		Employees are easily accessible when needed	0.620	0.569	
		Office is conveniently located	0.535	0.570	
	Availability	Benefits of the schemes are easily available	0.819	0.702	
		Benefits of the schemes are available within stipulated time limit	0.810	0.683	
		Procedure involved is easy	0.763	0.644	
		Employees follow quick problem-solving approach	0.601	0.579	
		Employees are helpful in making information available regarding the schemes	0.783	0.681	
	Usage	Schemes are successfully implemented	0.808	0.710	
		Other family members are also the beneficiaries of these schemes	0.732	0.611	
		Purchasing power has increased	0.724	0.670	
		Area has developed after implementation	0.629	0.688	

(*continued*)

Table 5.1 (continued)

Construct	*Factors*	*Items*	*Factor loading value*	*Communality value*	*KMO value*
Economic development	*Poverty reduction*	Household income has increased	0.735	0.683	0.723
		Sufficient money to spend on food grains	0.784	0.643	
		Started saving money for household	0.653	0.680	
		Payment of debt in time	0.723	0.671	
	Employment generation	Improvement in overall working conditions	0.748	0.563	
		No need to go out for employment	0.647	0.578	
		Provided diverse employment opportunities	0.628	0.629	
		Number of working days has increased	0.708	0.712	
	Standard of living	Per capita income of family has increased	0.824	0.731	
		Increase in value of assets	0.801	0.711	
		Less level of stress in life	0.746	0.691	
		Spend more on children's educational expenses	0.721	0.602	
		Empowered the family members	0.682	0.650	
Social development	*Social status*	Social status has increased over time	0.677	0.546	0.759
		Equal status for male and female workers	0.665	0.584	
		Benefits of the schemes have made people socially more reputed	0.631	0.501	
		Socially more developed after being covered under the schemes	0.533	0.527	
		Social relations have improved	0.528	0.576	
	Empowerment	Women of the families started working outside	0.727	0.650	
		Family members are accepted as members of the society	0.720	0.642	
		Unique id of family members created in the society	0.778	0.695	
		Reduced family crisis and social violence	0.785	0.635	
	Quality of life	Changed personality and lifestyle	0.814	0.680	
		Improved hygiene and education	0.635	0.495	
		Confidence level has increased over time	0.556	0.541	
		Contribution in bringing a change in the society	0.520	0.540	

emerged, namely, 'Poverty Reduction', 'Employment Generation' and 'Standard of Living'. The KMO value of 0.723 and BTS measure (chi-square = 620.825, df = 29 and p = 0.000) support the suitability of data for pursuing further analysis.

Social Development

The measure 'Social Development' contains 17 items, and it took 4 rounds to complete where 3 factors were extracted with 14 items having 5, 4 and 4 items in each factor. The KMO value came to be 0.759 and BTS values as chi-square = 541.321, df = 57 at 0.000 significant level with cumulative variance of 67.058 per cent. Further, the dimensions of the construct revealed good internal consistency having Cronbach's alpha value of 0.81.

Confirmatory Factor Analysis (CFA)

After EFA, confirmatory factor analysis was performed to assess fitness, reliability and validity of three separate models for government schemes, economic development and social development. Further, to assess the fitness of all three measurement models, the study used number of indices which include chi-square divided by degree of freedom (χ^2/df), root mean square error of approximation (RMSEA), normed fit index (NFI), comparative fit index (CFI), Tucker-Lewis index (TLI), relative fit index (RFI) and incremental fit index (IFI). All models produced satisfactory fitness results as all the fit indices satisfied the threshold criterion, that is, χ^2 value was less than 5.0; GFI, AGFI, CFI, TLI and NFI values were above or nearer to 0.90; and RMSEA and RMR values were less than the 0.08 (Table 5.2). During CFA, the items with SRW less than 0.5 were deleted (Hair et al. 2007). Second-order factor models have been framed for three scales as various factors emerged after applying EFA technique. First, second-order CFA was run to analyse government schemes model consisting of three measured factors, namely, accessibility, availability and usage with six, five and four items, respectively. Further, second-order CFA was carried on economic development construct with three measured factors. Finally, second-order CFA was run on social development construct with three factors, namely, 'Social status', 'Empowerment' and 'Standard of living' with five, four and four indicators, respectively. All three measurement models satisfied necessary conditions of identification as the fit indices of all the second-order models are within the prescribed limit (Table 5.2).

Table 5.2 Results of confirmatory factor analysis (CFA)

Construct	*Factors*	*Items*	*Standardised regression weight (SRW)*	*Model fitness*
Government schemes	*Accessibility*	4	0.79	**χ^2/df = 3.453,**
	Availability	3	0.63	**RMR = 0.049,**
	Usage	4	0.80	**GFI = 0.942, NFI = 0.969,**
				AGFI = 0.913, CFI = 0.978
				RMSEA = 0.061
Economic development	*Poverty reduction*	4	0.91	**χ^2/df = 1.996,**
	Employment generation	4	0.76	**RMR = 0.052,**
				GFI = 0.979, NFI = 0.872,
	Standard of living	3	0.62	**AGFI = 0.972, CFI = 0.942**
				RMSEA = 0.041
Social development	*Social status*	3	0.59	**χ^2/df = 2.986,**
	Empowerment	4	0.71	**RMR = 0.061,**
	Quality of life	3	0.68	**GFI = 0.981, NFI = 0.937,**
				AGFI = 0.936, CFI = 0.976
				RMSEA = 0.075

Reliability and Validity

To verify the internal consistency of the data, that is, the extent to which the scale used in the present study produces similar results over the period of time, Cronbach's alpha and composite reliability has been used as it indicates the reliability of three constructs (Hair et al. 2010). In the present study, the overall Cronbach's alpha value for all the constructs is found to be greater than 0.70 which is exhibited in Table 5.3. In addition to that, composite reliability for all the measures is found to be above 0.80 (Table 5.3). Therefore, the satisfactory values of Cronbach's alpha and composite reliability stated that three scales considered in the present study are reliable. Moreover, face and content validity of all the scales were also examined through existing literature reviewed and discussions with the government officials working in the Block Development Office (BDO), Udhampur. Further, convergent validity in the present study has also been established as SRW values of all scales are found to be greater than 0.5 and values of AVE (average variance explained) meet the acceptable criterion, that is, greater than 0.5 (Table 5.3). Finally, discriminant

Table 5.3 Reliability and validity results

Scales	*Alpha value*	*CR*	*AVE*	*Government schemes*	*Economic development*	*Social development*
Government schemes	0.890	0.95	*0.71*	*0.71*		
Economic development	0.903	0.88	*0.63*	0.222 0.471**	*0.63*	
Social development	0.810	0.93	*0.69*	0.312 0.559**	0.009 0.096*	*0.69*

Note: Values on the diagonal axis represent AVE, squared correlation is given below the diagonal axis and all the values within parentheses represent correlation. * = $p < 0.05$

validity of the scales has also been proved by comparing AVE with the squared correlation values among the constructs. In the present study, it has been found that AVE value of all scales is more than the squared correlation, hence confirmed the discriminant validity (Table 5.3).

7 Hypotheses Testing

After confirming the factors through CFA and confirming their validity and reliability, the next step performed is the application of structural equation modelling (SEM) to check the hypothesised relations. On the basis of SEM results, the framed hypotheses have been checked and the results are illustrated further.

Impact of Government Schemes on Economic Development

The first structural model is framed and the relationship between government schemes and economic development is checked. The effect of government schemes on economic development is examined in which model fit indices exhibit model to be good fit (χ^2/df = 1.878, RMR = 0.003, GFI = 0.977, CFI = 0.964, AGFI = 0.914, NFI = 0.930, RMSEA = 0.084). The SEM results clearly indicate that the government schemes significantly and positively affect economic development (SRW = 0.281, $p < 0.001$) (Fig. 5.2) and hence first hypotheses 'Government schemes significantly affect economic development' stands accepted.

Impact of Government Schemes on Social Development

Further, second SEM model has also been framed. In this model, the impact of all government schemes on social development has been checked. The result revealed that government schemes have been significantly affecting social development (SRW = 0.36, $p < 0.001$) (Fig. 5.2). Model fit indices are also good for this model (χ^2/df = 1.643, RMR = 0.006, GFI = 0.965, AGFI = 0.908, CFI = 0.963, RMSEA = 0.072) which lead to the acceptance of the second hypothesis 'Government schemes significantly affect social development'. Further, by examining and comparing the two hypothesised models, it can be observed that the SRW value of the second model (Fig. 5.3) is greater than the SRW of the first model (Fig. 5.1). Hence, it is indicated that the impact of government schemes on social development is greater than its effect on the economic development.

Impact of Government Schemes on Rural Development

Finally, SEM is used to test the relationship between government schemes and rural development as defined in the proposed research model. The SEM model fit indices, that is, model (χ^2/df = 2.298, GFI = 0.941, AGFI = 0.915, NFI = 0.903, CFI = 0.927, RMSEA = 0.068) reflects good model fit. The SEM results indicate that implementation of government

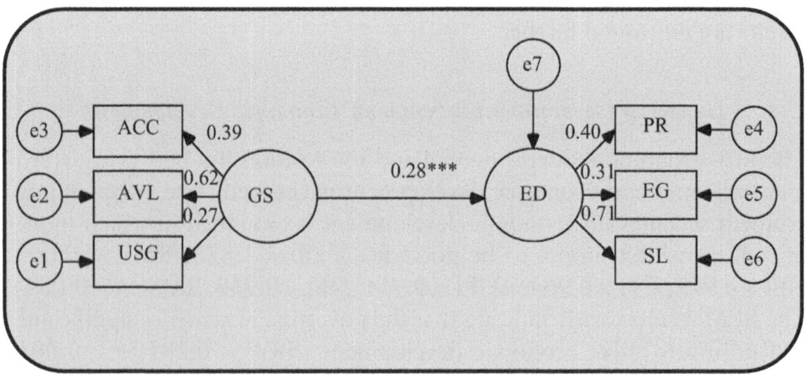

Fig. 5.2 Impact of government schemes on economic development. (Notes: GS, Government schemes; ED, Economic development; ACC, Acceptability; AVL, Availability; USG, Usage; PR, Poverty reduction; EG, Employment generation; SL, Standard of living are the observed variables and e1–e7 are the error terms)

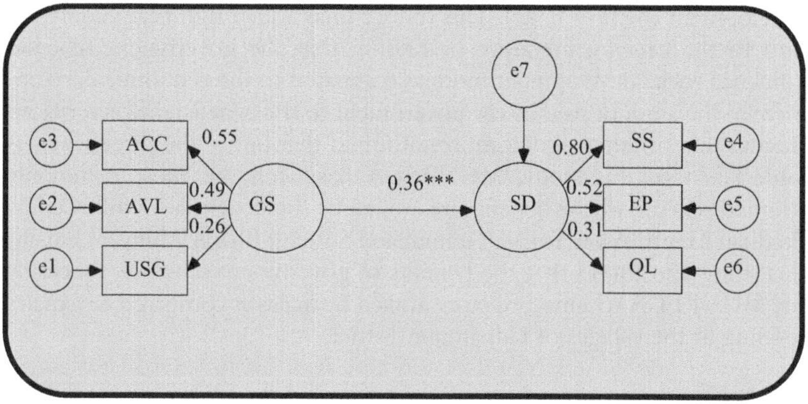

Fig. 5.3 Impact of government schemes on social development. (Notes: GS, Government schemes; ED, Economic development; ACC, Acceptability; AVL, Availability; USG, Usage; SS, Social status; EP, Empowerment; QL, Quality of life are the observed variables, and e1–e7 are the error terms)

schemes significantly influence rural development (SRW = 0.420, $p < 0.000$). Therefore, the final hypothesis, that is, 'Government schemes significantly affect rural development', is also accepted.

8 Discussion

The present study is conducted to have a comprehensive understanding of the implementation of various government schemes and their contribution towards the development of rural areas. The three schemes—which are extensively being run in the respective villages are assessed in the present research work—include MGNREGA, Pradhan Mantri Awas Yojana-Gramin and Swachh Bharat Mission. Moreover, the study also evaluates the role of these three schemes on economic and social development which are undertaken in Garnai and Malhar village of Udhampur district of Jammu and Kashmir. The empirical result of this study revealed that the government schemes significantly influence economic development (SRW = 0.28). Further, this study also found that the government schemes significantly and positively influence social development (SRW = 0.36). The study also examines the impact of government schemes on rural development, that is, social as well as economic development. The study results identified the significant and positive impact of the three government schemes on rural development (SRW = 0.42). This research has added significant contribution to the existing literature by finding that the government schemes influence social development more as compared to the economic development as the amount paid by the government to the beneficiaries is very low. Besides this, another significant result found that out of the three schemes undertaken in this study, MGNREGA has been the most commonly adopted scheme whose benefits are availed by the beneficiaries followed by Pradhan Mantri Awas Yojana-Gramin and Swachh Bharat Abhiyan. Finally, this study also found that the benefits of government schemes, especially the MGNREGA scheme, are more availed by males as compared to females residing in the villages of Udhampur district.

9 Implications

Policy Implications

Government has been introducing various schemes for rural development from time to time, but people for whom these schemes are meant many a time, remain unaware. Therefore, efforts should be made to improve awareness about various government schemes to the rural poor. Government should take more initiatives MGNREGA generally deals with those activities which do not require much skill; therefore, the need is to diversify the activities which need technical skills and abilities. It has also been found that the beneficiaries are not contented with the features of MGNREGA scheme as the working hours allowed under the scheme are very less. Moreover, beneficiaries are not satisfied with the number of working days allowed through this scheme and, therefore, it is appealed to the government to increase the number of working days and working hours. The beneficiaries are also dissatisfied with the amount of wages they get through MGNREGA; thus the beneficiaries want that the government should try to increase the wages provided under MGNREGA. The working environment in the schemes considered in the study is not satisfactory; therefore, there is a need to improve the working conditions while implementing the schemes.

Social Implications

Along with increase in economic development, the various rural development schemes undertaken in the study aim at social development of the rural poor. Therefore, government officials should take initiatives in providing proper education facilities, medical facilities and welfare activities for over all social development of the villagers. Further, Central and State Government should also take rigorous efforts by using proper channels to communicate and provide better social services in this district. Also, serious actions and supervision should be undertaken for implementing these schemes in each district of the state by appointing supervisor or competent authority so that regional disparities could be minimised.

10 Limitations and Future Research

The present study is confined to only three schemes, that is, MGNREGA, PMAY-G and Swachh Bharat Mission. Further, the study has examined the impact of government schemes on rural development, ignoring other outcomes like area development, and could also be checked as mediator. Moreover, this study examined the impact of government schemes on both economic and social development. Therefore, in future the mediating role of economic development between the relationship of government schemes and social development could be checked. Furthermore, the study is confined to the beneficiaries of two small villages and can be further extended across other areas of Udhampur area for the generalisability of results. The present study is cross-sectional and therefore, longitudinal studies should be conducted in future research work for more transparency of results. The presence of subjective responses of the beneficiaries with respect to schemes undertaken in the study is another limitation.

References

Ambasta, P. (2008). Two Years of NREGA: The Road Ahead. *Economic and Political Weekly, 43*(8), 41–50.

Arora, V., Kulshreshtha, L. R., & Upadhyay, V. (2013). Mahatma Gandhi National Rural Employment Guarantee Scheme: A Unique Scheme for Indian Rural Women. *International Journey of Economic, Practical and Theory, 3*(2), 14–18.

Bhargava, A. (2014). The Impact of India's Rural Employment Guarantee on Demand for Agricultural Technology. *IFPRI Discussion Paper 01381*. International Food Policy Research Institute.

Bhatty, K. (2006). Employment Guarantee and Child Rights. *Economic and Political Weekly, 41*(20), 1965–1967.

Burns, A. C., & Bush, R. F. (2007). Basic marketing research using microsoft excel data analysis. Prentice Hall Press.

Census. (2011). Retrieved from https://en.wikipedia.org/wiki/2011_Census_of_India

Das, D. (2012). Examining India's Mahatma Gandhi National Rural Employment Guarantee Act (MGNREGA): Its Impact and Women's Participation. *International Journal of Research in Management, 6*(2), 44–46.

Das, D. (2016). Role of MNREGA in Rural Employment-A Study of Arpeta District of Assam. *India'. International Journal of Humanities and Social Sciences, 2*(6), 241–248.

Datta, S. K., & Singh, K. (2012). Women's Job Participation in and Efficiency of MGNREGA Program-Case Study of a Poor District in India. *International Journal of Public Administration, 35*(7), 448–457.

De, L. C., Singh, D. R., Thapa, S., & Gurun, R. C. (2016). Swachh Bharat Abhiyan-An Overview. *International Journal of Information, Research and Review, 3*(11), 3066–3073.

Dodd, W., Wyngaarden, S., Humphries, S., Patel, K., Majowicz, S., Little, M., & Dewey, C. (2017). The Relationship Between MGNREGA and Internal Labour Migration in Tamil Nadu, India. *The European Journal of Development and Research, 30*(2), 1–17.

Dreze, J. (2005). Employment Guarantee Act-Promise and Demise. *Kurukshetra, 53*(7), 9–13.

Farooqi, S. A., & Saleem, I. (2015). Mahatma Gandhi National Rural Employment Guarantee Act (MGNREGA) and Empowerment of Women from BPL Families in Rural Areas. A Case Study of District Aligarh (India). *IOSR Journal of Humanities and Social Science (IOSR-JHSS), 20*(3), 7–16.

Hair, J. F., Black, W. C., Babin, B. J., & Anderson, R. E., & Tatham, R. L. (2007). Multivariate Data Analysis (6th ed.). New Jersey: Pearson Prentice Hall.

Hair, J. F., Black, W. C., Babin, B. J., & Anderson, R. E. (2010). Multivariate Data Analysis (7th ed.). New Jersey: Prentice Hall.

Jain, A. (2017). Social and Economic Impact of MGNREGA on Rural Workers: An Empirical Study of Problems and Prospects. *IOSR Journal of Humanities and Social Science, 2*(6), 38–43.

Karthika, K. T. (2015). Impact of MGNREGA on Socio-economic Development and Women Empowerment. *IOSR Journal of Business and Management, 17*(7), 16–19.

Krishnamurthy, J. (2006). Employment Guarantee and Crisis Response. *Economic and Political Weekly, 41*(9), 789–790.

Kumar, U., & Bhattacharya, P. (2013). Participation of Women in MGNREGA: How Far is it Successful in Morigaon, Assam. *Indian Journal of Economic and Development, 1*(2), 65–66.

Malhotra, N. K., & Dash, S. (2010). *Marketing Research: An Applied Approach*. Dorling Kindersely.

Manoj, P. K., & Alexon Puthukkeril, K. R. (2015). MGNREGS in Kerala: An Empirical Study of the Problems and Prospects with a Focus on Female Beneficiaries. *Indian Journal of Applied Research, 5*(10), 63–67.

Mathur, L. (2007). Employment Guarantee: Progress so far. *Economic and Political Weekly, 42*(52), 17–20.

Naskar, M., & Kumar, S. D. (2017). Pradhan Mantri Jan Dhan Yojana: An Empirical Study in Rural Area. *International Journal of Academic Research, 2*(6), 120–125.

Pamecha, S., & Sharma, I. (2015). Socio- Economic Impact of Mgnrega—A Study Undertaken among Beneficiaries of 20 Villages of Dungarpur District of Rajasthan. *International Journal of Scientific and Research Publications., 5*(1), 1–4.

Pandey, A. C., Soodan, V., & Bahuguna, R. (2016). A Study on Socio Economic Impact of MGNREGA on Beneficiaries in Rudraprayag District of Uttarakhand-India. *International Journal of Management and Applied Science, 2*(10), 44–52.

Patel, R., Patel, M., & Patel, N. (2018). Impact of microfinance on poor women: Lessons from North Gujarat. *Prabandhan: Indian Journal of Management, 11*(2), 14–29.

Poonia, J. (2012). Critical Study of MGNREGA: Impact and Women's Participation. *International Journal of Human Development and Management Science, 1*(1), 35–55.

Sami, L., & Khan, A. (2016). Mahatma Gandhi National Rural Employment Guarantee Act (MGNREGA): A Tool for Employment Generation. *International Journal of Social Science and Management, 3*(4), 281–286.

Shah, M. (2007). Employment Guarantee, Civil Society and Indian Democracy. *Economic and Political Weekly, 42*(45/46), 43–51.

Shaik, K. B. (2017). A Conceptual Analysis on Development of Rural Villages with Scarcely Resources in Prakasam District, Andhra Pradesh. *International Journal of Economic and Management, 6*(4), 1–4.

Shettar, R. M. (2016). Pradhan Mantri Jan Dhan Yojana: Issues and Challenges. *IOSR Journal of Business and Management, 18*(2), 17–24.

Socio-Economic Census. (2011). Retrieved from https://secc.gov.in/statewiseSeccDataSummaryReport

Sugapriyan, G., & Prakasam, S. (2015). Analysing the Performance of MGNREGA Scheme using Data Mining Technique. *International Journal of Computer Applications, 109*(9), 11–15.

Tiwari, N., & Upadhyay, R. (2012). Constraints Faced by the Women Beneficiaries under Mahatma Gandhi National Rural Employment Guarantee Act (MGNREGA). *Studies on Home and Community Science, 6*(2), 99–102.

CHAPTER 6

Agricultural Extension Agents and Challenges for Sustainable Development

Sunita Raina

1 Introduction

Background

The basic objective of agricultural extension agents (AEA) in any country is to overcome the challenges of food scarcity by introducing high yielding crops, improved cropping patterns and climate-compatible cropping systems as well as providing regular guidance to the farmers about pest management, storage management, storage techniques in addition to subsidising farmers' inputs and so on. Even small farmers are benefited through the work of AEA who provide regular guidance and advice regarding the benefits of newly improved agricultural technologies which consequently improves the productivity level as well as the overall socio-economic status of the rural population (Albore 2018; Raidimi and Kabiti 2017). India is an agricultural nation and its allied activities provide a source of livelihood to over 70 per cent of the total population. It occupies an important position in the Indian economy as its contribution in the gross domestic product (GDP) is about 30 per cent as compared to about 27 per cent from the

S. Raina (✉)
Government Degree College Kathua, Kathua, India

H. Chahal et al. (eds.), *Sustainable Business Practices for Rural Development*, https://doi.org/10.1007/978-981-13-9298-6_6

manufacturing and industrial sector. Agriculture provides food, fodder and raw material to agro-based industries particularly to those related to textiles, sugar, paper, vegetable and so on. Since majority of the farmers in India are smallholders whose subsistence and livelihood depend on agriculture (Singh et al. 2014a) by implementing new ways of farming, even small farmers can be in a position to capture their due market share and maximise their profits (Awotide et al. 2016).

Many researchers and practitioners have discussed that an efficient agricultural extension and information system is inevitable for boosting agricultural productivity in any country. The ground-level workers of both government bodies and NGOs play a significant role by providing regular support and guidance to the farmers at the field level to enhance their productivity levels which in turn lead to improvement of their socio-economic statuses. The extension services play a vital role in improving farmers' status by guiding and educating them about the new means of farming (Shahbaz and Ata 2014). No doubt AEA plays a significant role in the upliftment of farmers by helping them in using new high yielding cropping patterns.

But still the complete benefits are not yet achieved by all groups of farmers because of the differences in the socio-economic statuses and farm sizes. So an important task of the AEAs is to focus on these socio-economic and environmental factors before developing as well as transferring new ways of farming (Mohammad 2009). Recognising the contribution of agriculture to the overall socio-economic development of the country, new technologies for better farming are generated regularly at various agriculture research stations and agriculture universities such as Sher-e-Kashmir University of Agricultural Sciences and Technology (SKUAST) and Krishi Vigyan Kendra (KVK) (Shahbaz and Ata 2014; Veerasamy et al. 1994). The technology transfer and its popularization are indeed very crucial for translating the knowledge potential into a reality. Therefore, importance of right message at the right time and in the right manner has become more imperative because there is still a huge gap between what the farmers know and what they practice in their farms (Singh et al. 2014b; Bhagat et al. 2002; Bihari and Mishra 2001), which subsequently results in their low agricultural productivity.

Independent research by the present author (Raina 2008) validated that generally the new technologies are either not disseminated to the farming community or if disseminated are not properly received, understood or adopted by the farmers. Various demographic characteristics of farmers (Jahagirdar and Sundaraswamy 2002), poor functioning of extension agencies with respect to top-down approaches (Das and Saha 2002), small size of holding, scattered fields and inequality in ownership (Singh et al. 2014a), improper marketing facilities and casual attitude of extension professional, lack of adequate information system to keep extension agents up-to-date on new technologies are found to be some of the important reasons for ineffective implementation of the information.

Though over the years, a number of extension agencies such as Community Development Programmes (1952), National Extension Services (1953), Intensive Agriculture Area Programme (1964–1965), High Yielding Varieties Programme (1966–1967), Farmers' Training and Education Programme (1966–1967), Agriculture Development Programme (1971), Operating Research Programme (1975), Small and Marginal Farmers Development Programme (1969–1970), Training and Visit System (1974) and Krishi Vigyan Kendra (1974), Agricultural Technology Management Agency (1998–2005), Agricultural Technology Information Centers (2000) and so on, were established by the government, but still the scenario of agriculture is not properly developed in India (Kedia 2003; Sajesh and Suresh 2016; Singh et al. 2013, 2014b). Agricultural extension system provides all basic inputs in the form of subsidies, information, pesticides and other support that they need in the successful implementation of new technology enhancing their yield and food security. But it has been noticed that only selected new agricultural technology is adopted by majority of the farmers. This may be because the target consumers are aware of only a part of the technology generated at various agriculture stations and from this very limited new technology is accepted and utilised by them in their fields. In addition to this, Indian farmers are also under undue pressure to meet the food requirements of the continuously increasing population on the one hand and to compete with global agriculture market on the other. This necessitates the core providers of technology to farmers to be abreast of latest technology to ensure effective technology transfer for sustainable and overall development of the nation.

Jammu and Kashmir Scenario

The state of Jammu and Kashmir is mainly an agrarian economy where majority of its population depends on agriculture and allied sectors which include all small and medium enterprises that depend directly or indirectly on agriculture such as agricultural products, livestock, forestry logging and fishery. Agriculture and allied sectors contribute around 27 per cent to the state's income. Within the state, the Jammu division itself is home to a large diversity of physiographic features. Basmati rice and Rajmash (an important pulse crop) are valuable cash crops of the region. Vegetables, oil seeds, spices and fodder are also grown in specific areas of the region (Sharma 2018). The main thrust of the government focuses on adopting improved farming techniques for enhancing productivity level as well to fulfil the needs of growing population from the limited agricultural land available. The government in this regard has recognised a number of organisations such as Training and Visit System, Charcha Mandal, Krishi Vigyan Kendra, Agriculture Technology Management Agency and other NGOs which deliver the latest know-how techniques to farmers. AEA acts as the link between the agricultural research and farmers, as they acquire relevant scientific information from researchers and transfer it to the farmers on one hand and, on the other hand, they forward the farmer's problems to the agriculture scientists. For long these practices have been in use to exchange the production and productivity of agriculture necessary for the overall development of the farmers. The research-extension-farmers per se are considered the cardinal pillars for the implementation of participatory methodology at the field level (Kumar and Sah 2002). To harness the full benefits of new technologies, many reforms were taken in the Indian extension system from time to time, but still the expected results have not been achieved and there exists a wide transfer of technology (TOT) chasm. To bridge the gap between the technology generation station and the technology adoption station, due care should be taken for the availability of the basic inputs necessary for the implementation of new technology for sustainable development (Davidson et al. 2001). There is a dire need to develop the appropriate, location-specific and economically viable technology that would be compatible and suitable for the resource-poor farmer. Keeping in view the role of agriculture, extension agents and the importance of effective transfer of technology process (TOTP) to the farmers, the study was conducted in Jammu District.

2 Research Design and Methodology

Objectives

(a) To measure the overall effectiveness of the transfer of technology TOTP for sustainable development of the farmers.
(b) To find out major factors affecting TOTP vis-à-vis the technology gap that affects farmers' productivity and overall sustainability.
(c) To suggest strategic measures from extension personnel agents and perspectives to strengthening the TOTP for sustainable development of farmers and rural development.

Hypothesis

Agricultural technologies are being generated at different research stations and agriculture universities to enhance agricultural productivity vis-à-vis farmer's productivity (Slathia et al. 2012). However, the maximum benefits of technologies generated can be availed only when they are adopted by the farmers at their field levels (Illasaria and Soni 1993 and 1994; Patel et al. 2003). This mainly depends upon the AEA who are responsible for carrying the new technology from the agriculture scientists and transfer the same to the farmers. They help the tiller of the soil in their decision-making process at the field level and motivate them for the implementation of new farm technologies to enhance their productivity vis-à-vis sustainability. AEA can make extension services more effective by transferring up-to-date information to the farmers (Kassa 2016). Therefore, various governmental and NGO extension agencies are making continuous efforts to transfer technical know-how to the farmers (Directorate of Agriculture 2003). Among the governmental organisations, Training and Visit Extension System has created a breakthrough in disseminating the technologies to the farming communities in a planned and systematic manner through technically trained AEA. Under the system, the information passes through different stages, that is, from subject matter specialists (SMSs) to agricultural extension officers (AEOs) to junior agriculture assistants (JAAs) to Contact Farmers (CFs), and from CF finally the messages reach the Non-Contact Farmers (NCFs) who are the last users in the technology transfer channel stages (Directorate of Agriculture 2003; Venugopalan and Perumal 1992). Consequently there is an inevitable loss of message resulting in the big gap between the actual yield and potential yield (Thakur and Sinha 1989).

There exists a significant information loss during the transfer of technology process, particularly when the information is transferred from the extension personnel to the farmers.

Data Collection

The Jammu region comprises the districts of Kathua, Jammu, Udhampur, Doda, Rajouri and Poonch. The selection of AEA for the study purpose was done in three stages. At the first stage included two broader areas, that is, command area (CA) and non-command area (NCA) of Jammu District. CA refers to those areas which are mainly dependent on the Department of Agriculture for irrigation purposes like channels, drains and so on, whereas

NCA meets its irrigation requirement by its own ways such as tube well, hand pumps and so on. At the second stage, two blocks (Samba and Vijaypur) from command area were selected on census bases, and two blocks (R.S. Pura and Marh) from non-command area were selected on the basis of the recommendations made by the agriculture experts. In the second stage, a list of agricultural extension agents comprising 146 persons with 11 SMSs, 16 AEOs and 114 JAAs were procured from the Department of Agriculture. Lastly, the self-developed tested structure questionnaire was distributed to all the 146 extension agents. One hundred and forty-one out of 146 questionnaires were received that amounts to 97 per cent response rate. The final 141 respondents include 11 SMSs, 16 AEOs and 114 JAAs. The demographic profile of the 141 respondents is given in Table 6.1.

Reliability and Validity

Reliability is a measurement tool that measures the quality of an instrument by signifying the degree of variable errors present in the scale. The internal consistency and split-half reliability methods are used to assess the reliability of the scale (Chand and Sohal 1983; Hair et al. 2003; Malhotra 2002; Tull and Hawkins 1998). The internal consistency of the scale was measured using Cronbach's alpha values (CAV) for three selected variables, namely, 'Farmers', 'New Technology (NT)' and 'Job Satisfaction

Table 6.1 Demographic profile of agricultural extension personnel

Demographic characteristics	*Groups*	*Number*	*Percentage*
Gender	Male	135	95.7
	Female	6	4.3
Age (AG)	25–35 years	28	19.9
	35–45 years	51	36.2
	45–55 years	57	40.4
	Above 55 years	5	3.5
Education (ED)	B.Sc.	10	75.9
	M.Sc.	25	17.7
	Any other	9	6.4
Position (PO)	SMS	11	7.8
	AEO	16	11.3
	JAA	114	80.9
Experience (EX)	Below 5 years	19	13.5
	5–10 years	21	14.9
	10–15 years	89	63.1
	Above 15 years	12	8.5

Table 6.2 Variable-wise Cronbach's alpha, percentage of variance, Kaiser-Meyer-Olkin values along with Bartlett's Test of Sphericity of agricultural extension agents

Variables	*Cronbach's alpha values*				*Variance explained (%)*	*Kaiser-Meyer-Olkin*	*Bartlett's Test of Sphericity*	
	Sample	*Sub-Samples*					*Approx. chi-square values*	p
		I	*II*	*III*				
Farmers	0.55	0.64	0.57	0.49	64.72	0.61	196.18	0.000
New technology	0.55	0.33	0.66	0.71	60.15	0.60	184.84	0.000
Job satisfaction	0.67	0.60	0.64	0.74	73.75	0.61	234.92	0.000

(JS)'. The CAV for the overall sample is found to be highest for the variable 'JS' followed by 'Farmers' and finally by the variable 'NT' (Table 6.2). Further the sample of 141 respondents was divided into 3 equal parts, that is, sample I (1–47), sub-sample II (48–94) and sub-sample III (95–141) to determine the reliability. The CAVs of the sub-sample I were found to be 0.64, 0.33 and 0.60 for the 'Farmers', 'NT' and 'JS', respectively. Furthermore, under the sub-sample II, the variable 'NT' scored the highest CAV, and this was followed by 'JS' and 'Farmers'; the values scored were 0.66, 0.65 and 0.57, respectively. Under sub-sample III, two variables scored CAV above 0.7, that is, 0.74 for 'JS' and 0.71 for 'NT' variable, while the variable 'Farmers' scored low CAV, that is, less than 0.5, that is, 0.49 (Table 6.2). In addition to reliability, the validity of the instrument used for strengthening the credence of the study was also checked. The face, content and construct validity of the sample were assessed. Modifications were made in the questionnaire to make statements understandable to the respondents, which supported the face validity. The content validity was checked after thorough and threadbare discussions with AEAs, farmers and subject experts and review of literature. After the systematic evaluation of the content of the questionnaire, the construct validity was checked out using factor analysis on three variables, namely, 'Farmer', 'NT' and 'JS'. The KMO and percentage of variance were found to be 0.61 and 64.72 per cent, 0.63 and 56.97 per cent and 0.61 and 73.75 per cent for the three respective variables which supported the construct validity of the scale. Further, majority of the factor statements scored high factor loading values (FLVs) under three factors which consequently supported the construct validity of the scale.

3 Data Analysis and Findings

The relative impact of the three variables on overall effectiveness of TOT was analysed using factor analysis. The significant factors underlying the three selected variables are discussed in the following sub-sections.

Attitude of AEAs Towards Farmers

The factor analysis with principal component method was used to reduce and identify the most relevant items for the variable 'Farmers'. The KMO value was found to be (0.61), which specified the relevance of data for factors' identification and summarization. Five factors named as Selection of CF, Active Participation of CF, Role of CF in Effective TOTP, Varied Facilitating Functions and Lack of Participation by NCF with FLVs ranging between 0.54 and 0.84 (Table 6.3) are analysed further.

Table 6.3 Factor loading, communality, percentage of variance, Eigen, mean score and grand mean score values of agricultural extension agents towards 'Farmer Variable'

Item no.	*Items*	*FLV*	*CV*	*MSV*	*SA*	*A*	*I*	*D*	*SD*
F 1: Selection of CF (GMSV = 2.42, Eigen value = 1.82, percentage of variance = 15.18)									
10	Farmers' age	0.82	0.70	2.66	2.8	29.8	5.7	44.7	11.3
9	Farmers' education	0.74	0.63	2.49	2.8	22.7	9.2	40.4	17.7
8	Farmers' resources	0.54	0.46	2.11	1.4	16.3	2.8	41.8	29.8
F 2: Active Participation of CF (GMSV = 3.82, Eigen value = 1.73, % of variance = 14.41)									
7	CF desire for training	0.77	0.62	3.80	18.4	55.3	5.7	12.8	2.1
3	Feedback about problems	0.67	0.68	3.83	22.0	53.2	7.8	9.9	3.5
F 3: Role of CF in effective TOTP (GMSV = 3.82, Eigen value = 1.48, % of variance = 12.36)									
13	Feedback about TOT	0.80	0.72	3.50	9.2	54.6	8.5	12.8	7.1
1	Learning orientation	0.63	0.55	4.09	32.6	53.9	2.1	5.7	3.5
F 4: Varied Facilitating Functions (GMSV = 2.28, Eigen value = 1.45, % of variance = 11.96)									
4	Subsidies to CF and NCF	0.84	0.82	3.93	28.4	49.6	2.8	9.9	4.3
5	Equal TP for CF and NCF	0.70	0.83	3.51	22.7	39.0	5.0	24.1	5.7
F 5: Lack of Participation by NCF (GMSV = 3.29, Eigen value = 1.30, % of variance = 10.8)									
2	Implementation of NT	0.76	0.60	2.96	6.4	31.2	13.5	34.8	6.4
12	Learning orientation	0.64	0.60	3.62	13.5	56.7	4.3	16.3	4.3

Notes: F = Factor, FLV = Factor Loading Values, CV = Communality Values, MSV = Mean Score Values, GMSV = Grand Mean Score Values, SA = Strongly Agree, A = Agree, I = Indifferent, D = Disagree, SD = Strongly Disagree, CF = Contact Farmers, NCF = Non-Contact Farmers, TP = Training Programmes, TOTP = Transfer of Technology Process, NT = New Technology

Selection of CF

The extension system is based on the assumption that with limited staff at the ground level, all farmers cannot be contacted. The group of farmers, commonly known as CF, are selected for message dissemination purposes, who communicate the learned technology further to other fellow contact farmers (FCFs) and other NCFs (Dhama and Bhatnagar 1987). Thus, the success of the effective TOT greatly depends on the CF, and it is important that proper care is given while selecting the CF. The selection of CF involves two high FLVs for items, namely, 'farmer's age (FLV = 0.82)'and 'farmer's education (FLV = 0.74)' and one average FLV item 'farmer's resources (FL = 0.54)'. The high FLV indicates that the age and education are significant for the selection of CF, but at the same time 44.7 per cent and 40.4 per cent AEA, respectively, indicated that demographic factors particularly age and education were not generally taken into consideration for the selection of the CF (Table 6.3). The income factor is found not to be so relevant (Factor Loading Values, FLV = 0.54, and Mean Score Values, MSV = 2.11) for the selection of CF, but the study conducted by Davidson et al. (2001) for effective TOTP emphasised that the selection of CF is biased towards the large resource-rich farmers and those better positioned to adopt new technologies. The study conducted by Sanoria and Sharma (1983) in Sehore District of Madhya Pradesh revealed that T & V agency considered socio-economic conditions and progressive nature of farmers for the selection of CF. The MSVs for the two items, that is, farmer's age (MSV = 2.66) and farmer's education (MSV = 2.49), indicate the opposite. Amin and Stewart (1994) also found education and income to be a significant factor, but they found age to be an insignificant factor for the selection of the CF.

Active Participation of CF

The high FLV is scored for the first dimension, namely, 'demand for organisation of training (FLV = 0.77)', and average FLV for the second item, namely, 'regular feedback about field problems (FLV = 0.67)'. However, both items scored above average MSVs of 3.80 and 3.83, respectively. These findings reveal that CFs give regular feedback about their field problems and convey the NCF problems to the AEA. Around 55.3 per cent AEA reported that farmers were aware about their problems and demanded regular organisation of training programmes. Meena et al. (2002) also mention the active participation of the CF in adoption of recommended technology. Communality values (CVs) of the study further suggest that extension agents should have close liaison with this group of farmers to spread their message more quickly and effectively to the whole farming community. Further, CVs, that is, 0.62 and 0.68, respectively, for the items suggest the average degree of explanation power of these statements in understanding the impact of the role performed by CF in espousal of the new agro-techniques.

Role of CF in Effective TOTP

The third factor named as 'Role of CF in Effective TOTP' has also scored one high and one average FLV for 'information dissemination to NCF (FLV = 0.80 & MSV = 3.50)' and average 'learning orientation of CF (FLV = 0.63 and MSV = 4.09)'. AEAs state that CFs are more learning oriented towards NT and the same findings were also stated by Singh et al. (1989). Majority of AEAs (54.6 per cent) found that CF regularly disseminates the information to the NCF, the last channel in the TOT chain of T and V system. The study results contradict the results of studies by Davidson et al. (2001) where the CVs of 0.72 and 0.55, respectively, for the items indicate interrelationship between the two items within the 'Role of CF' factor.

Varied Facilitating Function

The main thrust of extension agency is to provide farmers with timely and relevant technical advice formally to the CFs and informally to all other farmers. The fourth factor of the 'Farmer's variable' in this regard helps to find out whether there exists any sort of discrepancy in providing varied services to the CF and NCF by AEA. The high FLV and CV, that is, FLV = 0.84 and CV = 0.82 and FLV = 0.70 and CV = 0.83 for the items, namely, 'equal subsidies to CF as well as NCF' and 'equal training for CF and NCF', respectively, showed the relevance of the dimension for TOT. Further, the above average MSVs of 3.93 and 3.51 for both the items (Table 6.3) indicate that presently AEA offers extension services equally to the CF and NCF.

Lack of Effective Implementation by NCF

Table 6.3 shows CV of the two items, namely, 'NCF are interested NT implementation (CV = 0.60)' and 'learning orientation of NCF (CV = 0.60)' which indicate the average level of internal homogeneity and variance of the two items, grouped under 'Lack of Effective Implementation by NCF' factor. Both the items are also scored mixed type of FLV and

MSV, that is, FLV = 0.76 and MSV = 2.96 and FLV = 0.64 and MSV = 3.62 for the respective items. The results, therefore, highlight that although NCF are interested in collecting the information from the AEA, they showed lesser amount of interest in adoption of the learned technology at their field levels. Around 56.7 per cent AEA also found the same. The study conducted by Singh et al. (1989) in four districts of Nepal also support these findings, but the study conducted in three administrative regions in Minya Governorate of Egypt by Amin and Stewart (1994) found that both groups are equally interested in implementing NT.

Attitude of AEA Towards New Technology (NT)

The 'NT' variable has 17 items and these were grouped into 4 important factors. The KMO value after the factor was found to be 0.60 and total percentage of variance as 60.15 and Bartlett's Test of Sphericity (BTS) measure as $X^2 = 184.84$ ($p = 0.000$). High FLVs were scored in the majority of the statements that were grouped into five different factors named as Basis for Technology Generation (five items), Customised Technology (two items), Influence of Age and Farm Size (two items) and NT (two items). These are further described as underneath:

Basis for Technology Generation

The CVs 0.56, 0.60, 0.58 and 0.69 for items 'alternative technologies is developed in consideration with the socio-economic condition of the farmers', 'agriculture scientists develops the NT according to the field problems of the farmers', 'sufficient infrastructures available for effective TOT' and 'modify CT according to the nature of NT used', respectively, show close association of all items with the 'Basis for Technology Generation'. The FLV and MSV were arrived at FLV = 0.72 and MSV = 3.06, FLV = 0.65 and MSV = 3.41, FLV = 0.63 and MSV = 2.95 and FLV = 0.50 and MSV = 3.82 for respective items. The results, therefore, acknowledged that the present extension system provided alternative technologies to different groups of farmers to increase the technology adoption rate. About 67.4 per cent extension agents stated that for getting the positive response from the farmers, they even modified the CT in accordance to the NT selected for TOTP. In addition, 56 per cent extension officers assured that the agriculture scientists take the farmer's needs and requirements into consideration while developing the technology. But the results based on the opinions of different categories of AEA, CF and NCF which were carried out by Kher and Patel (2000) in South Gujarat found a gap between the technology potential and actual yield. Such findings were earlier reported by Babu and Sinha (1979) as well. Similarly, Raina (2008) found that 40.3 per cent AEAs were dissatisfied with the availability of the sufficient infrastructure facility for TOT (Table 6.4). Chidi et al. (2015) also identified lack of proper infrastructure facilities in addition to economic problems and unfavourable government policies as the major constraints in effective TOTP.

Customised Technology

The second factor was christened as Customised Technology which comprised two items, namely, 'TOTP is need based (FLV = 0.82)' and 'NT is developed in consideration to the availability of the basic input resources (FLV = 0.69)', highlighted the role of these two items for technology generation process to ensure the adoption of the same in their fields by the target users. Hawley and co-workers (Howley et al. 2012) also acknowledged the positive association between farmers' involvement in farm advisory programmes and farmers' adoption process. Around 44.7 per cent AEA reported that the present TOT system does not ignore the farmers' requirement while developing and dissemination of the NT to them with above average MSVs of 3.38 and 3.56 for the two items (Table 6.4), respectively.

Influence of Age and Farm Size

The influence of age (CV = 0.65) and influence of farm size (CV = 0.67) indicated a high degree of the internal association of the two items within this factor. Both the items scored high FLV and moderately average MSV, that is, FLV = 0.77 and MSV = 3.05 and FLV = 0.80 and MSV = 3.03 for 'aged farmers' and for 'marginal farmers', respectively. AEA opined that age and farm sizes did not influence the farmer's decision to implement the NT. Similar results are observed by Chidi et al. (2015), Subashini and Thyagarajan (2002) and Kher (1992), but contrasting results were observed by Shrivastava et al. (2003) and Vijayaragavan and Subramanyan (1981). Subashini and Thyagarajan (2002) and Singh (1983) found significant association of land holding with the adoption of the NT.

New Technologies

The last factor, named as New Technologies contained two items, namely, 'educated farmers' and 'extension-research linkage', both the items scored high FLV and MSV, that is, FLV = 0.74 and MSV = 4.13 and FLV = 0.70

Table 6.4 Factor loading, communality, percentage of variance, Eigen, mean score and grand mean score values of agricultural extension agents towards 'New Technology Variable'

Item no.	*Items*	*FLV*	*CV*	*MSV*	*SA*	*A*	*I*	*D*	*SD*
F 1: Basis for NT Generation (GMSV = 3.31, Eigen value = 1.87, % of variance = 17.03)									
10	Social-economic factors	0.72	0.56	3.06	67.4	15.3	5.3	20.7	44.0
12	Field problems	0.65	0.60	3.41	56.0	12.7	12.7	20.7	34.7
8	Sufficient infrastructure	0.63	0.58	2.95	7.3	38.7	.7	28.0	40.3
7	Modify CT	0.50	0.69	3.82	12.0	16.0	13.3	31.3	8.7
F 2: Customized Technology (GMSV = 1.54, Eigen value = 1.66, % of variance = 15.13)									
1	Farmers' needs	0.82	0.71	3.38	44.7	17.3	3.3	21.3	3.3
2	Farmers' resources	0.69	0.53	3.56	13.3	12.7	11.3	10.0	0.7
F 3: Influence of Age and Farm Size (GMSV = 2.94, Eigen value = 1.57, % of variance = 14.24)									
17	Aged farmers	0.77	0.65	3.05	13.3	12.7	11.3	10.8	0.7
18	Marginal farmers	0.80	0.67	3.03	14.7	16.7	9.3	8.0	0.7
F 4: New Technologies (GMSV = 3.90, Eigen value = 1.51, % of variance = 13.75)									
19	Educated farmers	0.74	0.60	4.13	1.3	6.7	10.7	17.3	12.7
6	Feedback to scientists	0.70	0.58	3.67	4.7	12.7	1.3	18.7	52.7

Notes: F = Factor, FLV = Factor Loading Values, CV = Communality Values, MSV = Mean Score Values, GMSV = Grand Mean Score Values, SA = Strongly Agree, A = Agree, I = Indifferent, D = Disagree, SD = Strongly Disagree, CF = Contact Farmers, NCF = Non-Contact Farmers, TP = Training Programmes, TOTP = Transfer of Technology Process, NT = New Technology

and MSV = 3.67, respectively. The high FLV indicated their importance for effective technology generation and technology adoption process. The above average MSV along with high FLV indicates that educated farmers were more interested in using the latest information. Even the extension-research linkage was considered as quite significant by AEA. The results were also supported by Kher (1992) and Vijayaragavan and Subramanyan (1981). In addition, the results supported the role of the present extension system in carrying and disseminating the technology on the one hand and communicating the farmers' field problems to the agriculture scientists on the other. The CV 0.60 and 0.58 indicated the average explanation power of the items for the factor.

Attitude of AEAs Towards Job Satisfaction (JS)

The success of the TOTP process depends upon the level of JS of the AEA employed for the purpose of carrying and disseminating the NT from the research station to the farmer's field. This in turn depends upon the various socio-economic factors like age, experience, job commitment, rewards, salary, working environment and so on (Edward and Robert 1994; Ghosh and Vijayaragavan 2000; Reddy et al. 1992). In the present study, 'JS' variable scored variance of 73.75 per cent, KMO value of 0.61 and BTS ($\chi^2 = 234.92$, $p = 0.000$), whereas Singh et al. (2017) have found non-significant association of JS with age, gender, qualification, marital status and background. The factor analysis developed three important factors christened as Overall Job Satisfaction, Staff Relationships and Adequate Benefits. The FLVs are found to be ranged between 0.63 and 0.94. The brief portray of various factors is explained further.

Overall Job Satisfaction

The below average MSV for two important statements out of three selected statements, namely, 'scope for growth (MSV = 3.58)', 'administrative setup (MSV = 2.35)' and 'conducive working environment (MSV = 2.88)' with 0.80, 0.79 and 0.67 FLV, respectively, showed that AEAs are dissatisfied with the organisational setup and moderately satisfied with the overall organisational working environment (Ghosh and Vijayaragavan 2000; Kassa 2016). This might be the prime reason for the existing TOT gap. The CVs such as 0.69, 0.63 and 0.56, respectively, reflected moderate degree of association of the items with 'Overall Job Satisfaction'.

Staff Relationship

The items 'harmonious relationship (FLV = 0.93)' and 'cordial relationship (FLV = 0.94)' were identified in 'Staff Relationship' factor. The high FLV indicated the significance for developing harmonious relationship among staff personnel. About 66.7 per cent and 68.8 per cent AEAs are satisfied with their relations with their bosses and colleagues, respectively. Further the MSV for the respective items of 3.95 and 4.04 (Table 6.5) also ensured the effectiveness of technology dissemination process in the existing AEA. Ghosh and Vijayaragavan (2000) found that AEAs were averagely satisfied with the superior-subordinate relationship in the organisation. However, a recent study conducted by Kassa (2016) in Southwest Ethiopia has revealed that majority of extension personnel were not satisfied with the interpersonal relationships among co-workers and supervisors.

Adequate Benefits

Factor three of this variable scored CV and FLV of 0.80 and 0.88 and 0.73 and 0.63 against two items, namely, 'salary' and 'non-financial benefits',

Table 6.5 Factor loading, communality, percentage of variance, Eigen, mean score and grand mean score values of extension agents towards 'Job Satisfaction Variable'

Item no.	*Items*	*FLV*	*CV*	*MSV*	*SA*	*A*	*I*	*D*	*SD*
F 1: Overall Job Satisfaction (GMSV=2.94, Eigen value = 2.05, % of variance = 29.34)									
7	Scope for growth	0.80	0.69	3.58	16.3	54.6	3.5	7.1	12.8
6	Conducive environment	0.79	0.63	2.88	9.9	33.3	4.3	25.5	22.7
3	Administrative setup	0.67	0.56	2.35	3.5	20.6	8.5	33.3	27.7
F 2: Staff Relationship (GMSV = 24.00, Eigen value = 1.77, % of variance = 25.33)									
5	Cordially interrelationship	0.94	0.88	3.95	17.0	66.7	4.3	7.8	0.7
4	Harmonious behaviour	0.93	0.88	4.04	18.4	68.8	2.1	5.0	0.7
F 3: Adequate Benefits (GMSV = 32.81, Eigen value = 1.34, % of variance = 19.09)									
1	Salary	0.88	0.80	3.28	12.1	48.9	2.8	21.3	12.1
2	Non-financial benefits	0.63	0.73	2.33	2.1	24.1	4.3	35.5	27.7

Notes: F = Factor, FLV = Factor Loading Values, CV = Communality Values, MSV = Mean Score Values, GMSV = Grand Mean Score Values, SA = Strongly Agree, A = Agree, I = Indifferent, D = Disagree, SD = Strongly Disagree, CF = Contact Farmers, NCF = Non-Contact Farmers, TP = Training Programmes, TOTP = Transfer of Technology Process, NT = New Technology

respectively. The MSV of 2.33 for the item 'non-financial benefits' indicates that AEAs were not satisfied with the non-financial benefit incentives given to them (Table 6.5). However, AEA scored average satisfaction score for 'salary (MSV = 3.28)', whereas Ghosh and Vijayaragavan (2000) found contrasting results with respect to salary.

Hypotheses Testing

The study results identified significant loss of information during the TOTP which ultimately leads to low degree of farmers' productivity which was reflected from the MSVs of 3.41, 2.17 and 3.37 for the statements, namely, 'information loss at every stage of TOTP', 'information loss is more from extension to farmers', and 'information loss is less from research to extension' respectively. Also, the *t*-value was found to be 4.16 ($p = 0.000$) at 95 per cent confidence level that resulted in the acceptance of the first hypothesis regarding existence of significant loss of information at every stage of TOTP. Additionally, the *t*-values were calculated as −8.18 ($p = 0.000$) and 3.45 ($p = 0.010$), respectively, at 95 per cent confidence level, which confirmed more loss of information from extension to farmers in comparison to information loss from research to extension (Table 6.6).

Table 6.6 Mean, *t*- and *p*-values related to loss of information from research-extension-farmers

Hypotheses	*Test statistics*	*Test statistics*	t	*Mean*	p
Loss of information	One sample *t*-test (at test value = 3)	There exists information loss at every stage of technology process	4.158	3.41	0.000
		Information loss from extension personnel to farmers	−8.179	2.17	0.000
		Information loss from research station to extension personnel	3.452	3.37	0.010

Sig. at 0.05 level of significance (Raina 2008)

4 Problems

In a scenario of shrinking land and depleting water resources, the challenge of agriculture extension system is to increase biological yields of farmers to feed the ever-growing populating without destroying the ecological foundation of the soil. But it has been observed that generally modern technology did not match with the poor farmer's needs as well as resources such as availability of fertile land, labour, marketing facilities, irrigation facilities, timely seed, fertilisers, storage and so on and, therefore, farmers' participation is lacking in effective implementation of the new technology transferred by the extension agents. However, the study on the bases of low mean score values (Table 6.7) evolved out lack of proper selection of CF, less learning orientation of farmers, ineffective organisational functioning and lack of adequate infrastructure facilities as core problems which threaten the effective TOTP under existing agriculture extension system from AEA perspective and consequently have an effect on sustainability of the farmers and overall rural development.

Lack of Proper Selection of Contact Farmers

The studies found that extension agency did not take into consideration various socio-economic factors particularly age, education and resources while selecting the CF for TOT to other farmers. This was reflected by the MSVs of 2.66 for 'farmer's age', 2.49 for 'farmer's education' and 2.11 for 'farmer's resources' by 50 per cent of AEA. Further, the *t*-values which

Table 6.7 Barrier for effective transfer of technology process

Problems	*Problems*	*Items*	*Mean score values*	%	t	p
1	Lack of objective selection of contact farmers	Farmers' age	2.66	44.7	−3.44	0.000
		Farmers' education	2.49	40.4	−5.11	0.000
		Farmers' resources	2.11	41.8	−9.3	0.000
2	Less learning orientation of non-contact farmers	Effective implementation of technology by non-contact farmers	2.96	34.8	−0.39	0.700
3	Ineffective organizational functioning	Conducive working environment	2.88	33.3	17.2	0.000
		Administrative setup and policies	2.35	33.3	−6.12	0.000
		Non-financial benefits	2.33	35.5	−6.31	0.000
4	Lack of adequate infrastructure facilities	To provide timely information to farmers	2.95	40.3	−0.47	0.640
		To organise workshops	2.95	34.0	−0.43	0.670

were calculated as −3.44 ($p = 0.00$), −5.11 ($p = 0.00$) and −9.3 ($p = 0.00$) for the respective dimensions also strengthened the results.

Less Learning Orientation of NCFs

The MSV 2.96 ($t = -0.39$, $p = 0.70$) as scored by 34.9 per cent AEA for the dimension, namely, 'NCF are interested' in adopting 'NT', highlighted another major setback in the effectiveness of TOTP. But Amin and Stewart (1994) found the opposite and stated that both CF and NCF were equally interested in TOTP.

Ineffective Organisational Functioning

The MSVs of 2.35 ($t = -6.12$, $p = 0.00$), 2.99 ($t = 17.2$, $p = 0.00$) and 2.33 ($t = -6.31$, $p = 0.00$) for the dimensions, namely, 'administrative policies', 'working environment' and 'non-financial benefits', respectively, indicated that AEAs were not satisfied with the administrative policies and procedures of the organisation and as such perceived negatively about the overall working environment and provisions for non-financial benefit per se; these factors obstruct AEA performance in TOTP, leading to information chasm.

Lack of Adequate Infrastructure Facilities

The MSVs of 2.95 ($t = -0.47$, $p = 0.64$) and 2.95 ($t = -43$, $p = 0.67$) towards the statements, namely, 'inadequate infrastructure TOT to farmers' and 'inadequate infrastructure to effective monthly training' scored by the AEA (40.3 per cent) and (34.0 per cent), respectively, indicated that SMSs are not satisfied with the basic infrastructure required for organising varied print media for the trainer of trainers particularly the JAA as well as for the farmers. Thus, in absence of practical training for the JAA, very often JAAs were not in a position to solve farmer's field problems on the spot, and they depended fully on the SMS which ultimately caused unnecessary delay in effective adoption of NT by farmers. This further widens the TOT gap.

5 Suggestions

On the basis of the results discussed previously, the following important suggestions are proposed that ought to be incorporated by the existing extension agency to improve TOTP effectiveness.

Objective Selection of Contact Farmers

To make TOT an effective process, there is a dire need to concentrate on the root issue of selection of farmers. For this, effectual efforts are to be made to choose the lead farmers after duly taking into consideration the farmers' age, education, resources availability, interest social image and so on. This practice can help in selecting such progressive farmers who can guide, influence and control the thoughts, feeling and behaviour of not only other fellow farmers but also NCF. As a result, the adoption rate of technology learned from various levels of extension personnel vis-à-vis lead farmers can be increased eventually.

Information Diffusion to Non-Contact Farmers (NCFs)

For awarding and motivating majority of NCF groups, the AEA should also see that technology to be transferred is delivered effectively and timely to the NCF too. AEA can either directly communicate the necessary NT to

NCF or indirectly through CF, but for better results it is suggested to give more focus on direct transmission of the technology. Further, to approach all groups of farmers, particularly belonging to geographically widespread areas, the extension department should provide free transportation facilities to the field-level extension workers, that is, JAAs. Also, the indirect approach of contacting NCF through CF would be taken care of by incorporating 'objective selection of CF'. Furthermore, to provide maximum benefits of NT to NCF, regular and subsidised inputs supply must be ensured to this group of farmers particularly the resource-poor group.

Organising Regular Events for Farmers

The training programmes, educational tours and melas should be regularly organised within and outside nearby and progressive states such as Punjab and Haryana.

Post-Extension Services

The main motto of the present agricultural extension system remains confined only to awarding and motivating the farmers about new farming technology and implementing the same for enhancing their productivity level. It has limited role in post-harvesting and marketing of the agriculture produces. As a result, the farmers have to depend on the private local traders, mandis and mills for marketing their produce. This results in their low yearly income level even when their productivity has increased. Apart from providing regular pre-extension service, AEA should also provide equal consideration to the post-extension services such as processing, grading and packaging of their produce with assurance that their produce will reach the buyers directly and safely without any delay or loss and without dependence on intermediaries. This approach will subsequently improve farmers' bargaining position and can fetch them high returns. The AEA can meet this objective by opening shops mainly at the local level where farmers can easily sell their produce at reasonable prices and avail facilities such as transportation, cold storage, insurance in case of any damage and so on.

Resources Availability

Free input resources to the needy farmers and subsidised input resources particularly seeds and fertilisers to all contact and non-contact farmers should be offered by the extension wing of the Directorate of Agriculture, Jammu and Kashmir.

Development of Need-Based Technology

New technology should be developed after giving due consideration to the agro-climatic conditions and cropping patterns of the area (Nikam and Kumar 2017).

Growth and Development of Extension Personnel

To enhance job satisfaction of AEA which subsequently leads to effective TOT, it is suggested that due care should be given to the growth and development of the AEA. This can be achieved by providing equal promotional chance to everyone using merit-cum-seniority mixed strategies. Further, to boost their morale for better performance, the department should also keep provision for various financial and non-financial benefits like adequate residential accommodation, educational facilities for their children, departmental transportation facilities for approaching all groups of the plains as well as in kandi (dry areas that lack irrigation facilities) and hilly areas. Implementation of all these points can subsequently work for the development of conductive working environment for effective TOT.

Regular Training Programmes for the Trainers

To help the farmers to cope with the changing agricultural scenario at national as well as international levels, it is important that the extension services should be very fast in developing, carrying and disseminating the new farming techniques according to the farmer resources availability of farmers, their needs and requirement along with developing marketing opportunities for their produce. Adequate opportunity to AEAs in participating varied seminars, workshops, conferences, technological mela and so on can equip them with latest technology. The development of information resource centre can also provide sound collection of offline and online sources such as books, journals and newsletters published within and outside the nation which are useful for the skill enhancement of the AEA. Furthermore, arrangements should be made to provide 24 by 7 internet facilities to AEAs to not only access various e-journals, e-books and articles but also to contact different resource personnel from varied states and countries directly for solving and discussing agriculture related issues and technology.

Integrated Agriculture Information Systems

It is further suggested to establish intranet-based Integrated Agriculture Information System (IAIS) in the Directorate of Agriculture which can help to maintain regular links within the departments like input, state land use board and so on and also with other government organisations and NGOs working for the overall socio-economic upliftment of the farmers. In addition, sharing of varied information such as expenditure details of specific departments incurred for providing different financial and motivational benefits to employees, as well as the information regarding the development of new infrastructure, employment opportunities and new recruitments in the specific departments and so on, is also suggested. The information pertaining to farmers' requirement such as input distribution, yearly output, irrigated land, yield potential, the number of contacted and benefited farmers can also be maintained. Also, the establishment of IAIS can help in redesigning various policies of the department from holistic perspective which will be more effective for the overall growth and development of farmers and bridging TOT gap.

Inter-Coordination Among Agriculture Agencies

Lack of sufficient infrastructure facilities is considered to be the major obstacle in the TOT pathway. The government extension organisations should try to identify different sources/organisations which are working independently for farmers' growth and development for collaborative activities. In this regard it is suggested that both government and NGOs should come together to work for the effectiveness vis-à-vis farmers' productivity. In this context, the two big NGOs, namely, Alami Khudhi Khadmat Ghar and Paryavaran, working in varied agriculture-related activities such as watershed projects, mushroom cultivation, lady finger production and so on of Jammu region, should be considered for collaboration. In this context, both organisations should initiate joint plans and programmes regarding information collection and information diffusion process, provision of input facilities, training, marketing facilities and so on, to reach maximum number of farmers on the basis of their socio-economic and geographical requirements. This strategy will provide a pathway to fill up the existing TOT gap.

Women Empowerment

Women play a significant role in agricultural activities particularly related to post-harvest activities such as processing, preservation and utilisation of agriculture produce. Unfortunately, this group of rural population has abysmal access to the latest agriculture-related knowledge and new methodologies, and this hampers the growth and development of the farming families. Therefore, special attention is required to educate the women (Nikam and Kumar 2017) on various farming aspects using communication device such as radio, cassettes, television and so on which can help them in acquiring required information within their social, cultural and customary rituals/situations.

6 Managerial Implications

The implementation of aforesaid strategies relates to and stresses on certain financial and non-financial implications. The financial implications include organising training programmes for extension personnel and farmers, regular and timely promotional avenues and incentives to extension personnel to boost them for effective TOTP, to provide necessary subsidies on basic inputs, marketing facilities to the farmers to encourage them for implementing the new technologies. The funds for these activities could be covered by the government by allocating certain proportion of agriculture budget along with carrying activities in coordination with some NGOs. The non-financial implication on the other hand focuses on redesigning the selection procedure of farmers on the basis of the factors such as education, progressiveness and experience for technology dissemination purposes to the other fellow farmers and non-contact farmers, involvement of AEA in the core problems of the farmers, to organise training programmes for extension personnel and farmers, to provide regular supply of basic inputs and marketing facilities, and so on, to the farmers to encourage them for implementing the new technologies which subsequently can help in bridging existing technology gaps.

7 Conclusion

The extension workers are the last and the most important link in the chain connecting researchers and farmers. The role of agricultural extension in national development is relevant in achieving national food security, improving rural livelihood, empowering farmers by building social capital or improving natural resources management. The central government supports the Indian extension agencies for ensuring effective technology dissemination processes, but the proper transfer of technology process is still a big challenge for the Indian extension agencies which in turn affects the overall sustainability process in rural areas. This work has highlighted various issues which hinder the effectiveness of agricultural extension system. After independence, a series of rural development programmes have been undertaken to boost rural development strategies and efforts. However, unfortunately many of such programmes were stopped without performing a systematic assessment. Lack of support for proper implementation of these strategies at ground level is yet a big challenge (Shahbaz and Ata 2014). It is also equally important to take socio-economic statuses of farmers into consideration before generation and transmission of new technologies at various agriculture research stations. This chapter has also identified many intrinsic weaknesses of public extension services. For instance, modern technology did not match with the poor farmers' needs as well as resources such as availability of fertile land, labour, marketing facilities, irrigation facilities, timely seed, fertilisers, storage and so on, and, therefore, farmers' participation is lacking in effective implementation of the new technology transferred by the extension agents. However, the study evolved out lack of proper selection of CF, less learning orientation of NCF, ineffective organisational functioning and lack of adequate infrastructure facilities as core problems which threaten the effective TOTP. As such a complete participatory approach to agriculture development should be encouraged in new technology invention-dissemination for overall sustainable and rural development process of the state.

8 Limitations and Future Research

All feasible efforts have been made to maintain objectivity, validity and reliability of the study, but still the presence of subjectivity could not be ruled out. The study is restricted to the Jammu District only and as such can be extended to Jammu Province or different states to evaluate the agricultural extension services system for further research. The present study being AEA oriented has ignored the farmers' perspective regarding transfer of technology process. Thus dyad approach has to evolve out effective strategies for bridging TOT chasm.

References

Albore, A. (2018). Review on Role and Challenges of Agricultural Extension Service on Farm Productivity in Ethiopia. *International Journal of Agricultural Education and Extension, 4*(1), 93–100. Retrieved June 7, 2018, from www.premierpublishers.org.

Amin, A. H., & Stewart, B. R. (1994). Training and Visit (T&V) Extension Programme Impact in Minya Governorate, Egypt. *Journal of Indian Agricultural Extension, 35*(3), 30–34.

Awotide, B. A., Karimov, A. A., & Diagne, A. (2016). Agricultural Technology Adoption, Commercialization and Smallholder Rice Farmers' Welfare in Rural Nigeria. *Agricultural and Food Economics.* https://doi.org/10.1186/s40100-016-0047-8. Retrieved June 7, 2018.

Babu, A. R., & Sinha, B. P. (1979). Information Gap Between AEA and Farmers with Regard to Modern Rice Technology. *Indian Journal of Extension Education, 15*(3, 4), 52–59.

Bhagat, G. R., Nain, M. S., & Kher, S. K. (2002). Knowledge and Adoption Gap of Wheat Technology in Jammu. *Journal of Research, Sher–E-Kashmir University of Agricultural Sciences and Technology of Jammu, 1*(2), 192–196.

Bihari, B., & Mishra, A. S. (2001). Information Revolution in Indian Agricultural. *Manage Extension Research Review, 11*(2), 62–69.

Chand, R., & Sohal, T. S. (1983). A Comparison of Techniques for Measuring Aspiration of Dairy Farmers. *Indian Journal of Extension Education, 19*(1 & 2), 95–98.

Chidi, I., Anozie, R. O., & Priscilia, N. C. (2015). Analysis of Socio-economic Factors and Profitability of Rice Production Among Small Scale Farmers in Ebonyi State. *Journal of Agriculture and Veterinary Science, 8*(2), 20–27. Retrieved June 16, 2018, from www.iosrjournals.org.

Das, S., & Saha, B. (2002). Agriculture Technology Management Agency a New Approach of Technology Dissemination. *Kurukshetra, 50*(8), 34–37.

Davidson, A. P., Ahmad, M., & Ali, T. (2001). Dilemmas of Agricultural Extension in Pakistan: Food for Thought. *Agricultural Research and Extension Network Paper No. 116,* 1–21. Retrieved June 16, 2018, from https://www.researchgate.net/publication/242618776

Dhama, O. P., & Bhatnagar, O. P. (1987). *Education and Communication for Development* (pp. 38–79). Mohan Primlani for Oxford and IBH Publishing Company.

Directorate of Agriculture (DOA). 2003. *Report on Work Responsibilities of Extension Staff.* Department of Extension, Directorate of Jammu, Jammu, J&K.

Edward, N. S., & Robert, A. A. (1994). A Comparative Study of Management Effectiveness Under Training and Visit and General Extension in Ghana. *Journal of Indian Agricultural Extension, 35*(4), 36–41.

Ghosh, S., & Vijayaragavan, K. (2000). Performance Appraisal Climate and Job Satisfactions of Extension Personnel. *Indian Journal of Extension Education, 34*(3, 4), 85–91.

Hair, J., Jr., Rolph, E. A., Ronald, L. T., & William, C. B. (2003). *Multivariate Data Analysis with Readings*. New Jersey: Prentice Hall.

Howley, P., Donoghue, C. O., & Heanue, K. (2012). Factors Affecting Farmers' Adoption of Agricultural Innovations: A Panel Data Analysis of the Use of Artificial Insemination Among Dairy Farmers in Ireland. *Journal of Agricultural Science, 4*(6). Retrieved June 16, 2018, from www.ccsenet.org/jas.

Illasaria, N. M., & Soni, M. C. (1993 and 1994). Technology Gap Between Contact Farmers and Non-contact Farmers in Gram Production Technology. *Gujarat Journal of Extension Education, IV & V*, 190–192.

Jahagirdar, A. K., & Sundaraswamy, B. (2002). Adoption of Recommended Practices of Tomato Cultivation. *Agricultural Extension Review, 14*(3), 12–16.

Kassa, G. (2016). Determinants of Job Satisfaction Among Agricultural Extension Workers in Southwest Ethiopia. *American Journal of Agriculture and Forestry, 4*(5), 112–117. Retrieved June 15, 2018, from http://www.sciencepublishing-group.com/j/ajaf.

Kedia, O. (2003). Extension Services Key to Growth of Agriculture. *Kurukshetra, 51*(10), 43–45.

Kher, S. K. (1992). Adoption of Improved Wheat Cultivation Practices. *Indian Journal of Extension Education, 28*(1, 2), 97–99.

Kher, S. K., & Patel, R. B. (2000). A Study on Knowledge Gap at Various Levels of Technology Transfer with Respect to Paddy Crops. *Indian Journal of Extension Education, 34*(3, 4), 164–166.

Kumar, S., & Sah, U. (2002). Agricultural Extension for Rural Development Issues and Challenges. *Kurukshetra, 15*(4), 7, 19.

Malhotra, N. K. (2002). *Marketing Research: An Applied Orientation* (pp. 266–291). Singapore: Pearson Education.

Meena, R. P., Chandra, S., & Chaudhary, M. (2002). Adoption of Weather Forecast by Contact Farmers. *Agricultural Extension Review, 14*(3), 6–8, 16.

Mohammad, S. A. (2009). Agricultural Sustainability: Implications for Extension Systems. *African Journal of Agricultural Research, 4*(9), 781–786. Retrieved June 6, 2018, from http://www.academicjournals.org/ajar.

Nikam, V., & Kumar, S. (2017). Agriculture and Extension Policies in India: Connect and Disconnect with Nutrition. *Indian Journal of Nutrition, 4*(2). Retrieved June 8, 2018, from www.opensciencepublications.com/fulltextarticles/IJN-2395-2326-4-161.html.

Patel, M. M., Yadav, N. S., & Chatterjee, A. (2003). Technological Knowledge of Onion Growers. *Gujarat Journal of Extension Education, 14*, 1–4.

Raidimi, E. N., & Kabiti, H. M. (2017). Agricultural Extension, Research, and Development for Increased Food Security: The Need for Public-Private Sector

Partnerships in South Africa. *South African Journal of Agricultural Extension, 45*(1), 49–63. Retrieved June 6, 2018, from http://www.scielo.org.za/scielo.php?script=sci_serial&pid=0301-603X&lng=en&nrm=iso.
Raina, S. (2008). Significance of TOTP in Enhancing Farmers' Productivity. Thesis submitted to University of Jammu, Jammu.
Reddy, M. N., Reddy, P. G., & Das, U. C. (1992). Job Performance of Agricultural Extension Officers. *Indian Journal of Extension Education, 28*(3, 4), 87–89.
Sajesh, V. K., & Suresh, A. (2016). A Public-Sector Agricultural Extension in India: A Note. *The Review of Agrarian Studies, 6*(1), 1–16. Retrieved June 6, 2018, from http://www.ras.org.in/public_sector_agricultural_extension_in_india.
Sanoria, Y. C., & Sharma, D. K. (1983). Comparative Analyses of Adoption Behavior of Beneficiaries of Farm Development Programmes. *Indian Journal of Extension Education, 19*(1, 3), 84–86.
Shahbaz, B., & Ata, S. (2014). Agricultural Extension Services in Pakistan: Challenges, Constraints and Ways Forward. Retrieved June 23, 2018, from http://www.academia.edu
Sharma, Y. (2018). Agricultural Potential of Jammu and Kashmir. *News on Air*. Retrieved January 1, 2018, from www.newsonair.nic.in/Agricultural-potential-of-Jammu-and-Kashmir.asp
Shrivastava, K. K., Mohaniya, G. S., & Lakhera, M. L. (2003). Impact of Mushroom Training Programmes on Rural Women in Chhattisgarh State. *Gujarat Journal of Extension Education, 14*, 25–30.
Singh, A., Bhatti, J. S., Singh, J., & Sodhi, S. S. (2017). A Relationship of Socio-personal Factors with Job Satisfaction of Veterinary Officers in Punjab. *Journal of Animal Research, 7*(1), 123–127. Retrieved June 16, 2018, from https://www.researchgate.net/publication/316543253.
Singh, K. M., Meena, M. S., Singh, R. K. P., & Kumar, U. (2013). Agricultural Technology Management Agency (ATMA): A Study of Its Impact in Pilot Districts in Bihar, India. Retrieved January 1, 2018, from https://www.researchgate.net/publication/235975667_Agricultural_Technology_Management_Agency_ATMA_A_Study_of_its_Impact_in_Pilot_Districts_in_Bihar_India
Singh, K. M., Meena, M. S., Swanson, B. E., Reddy, M. N., & Bahal, R. (2014a). *In Depth Study of the Pluralistic Agricultural Extension System in India*. Retrieved June 20, 2018, from http://mpra.ub.uni-muenchen.de/59461
Singh, P., Vijay, D., Srinivas, K., & Wani, S. P. (2014b). Potential Productivity, Yield Gap, and Water Balance of Soybean-Chickpea Sequential Systematic Selected Benchmark Sites in India. Retrieved June 24, 2018, from https://www.researchgate.net/publication/26513627
Singh, R. (1983). Selected Characteristics of Farmers in Relation to Their Adoption of Farm Mechanization. *Indian Journal of Extension Education, 19*(3, 4), 11–17.

Singh, S. R., Singh, S., & Singh, K. B. (1989). Impact of Training and Visit System in the Adoption of Agricultural Technology in Nepal. *Maharashtra Journal of Extension Education, 8*, 287–289.

Slathia, P. S., Paul, N., Nain, M. S., Nanda, R., & Peshin, R. (2012). Credibility Crisis Among Agriculture Extension Functionaries in Jammu and Kashmir. *Indian Journal of Extension Education, 48*(1, 4), 68–73.

Subashini, B., & Thyagarajan, S. (2002). Socio-personal Characteristics of Farmers and Adoption of Tapioca Technology. *Indian Journal of Extension Education, 34*(3, 4), 153–115.

Thakur, K. K., & Sinha, R. R. (1989). Analysis of Agricultural Credit in Relation to Socio-economic Development of Farmers. *Maharashtra Journal of Extension Education, 8*, 119–122.

Tull, D. S., & Hawkins, D. I. (1998). *Marketing Research: Measurement and Methods*. New York: Macmillan Publishing.

Veerasamy, V., Satapathy, C., & Rao, A. (1994). Information Input, Processing and Output Behavior of Extension Personnel. *Indian Journal of Extension Education, 30*(1–4), 35–41.

Venugopalan, M., & Perumal, G. (1992). Applicability of Dry Land Technologies as Viewed by Extension Personnel. *Indian Journal of Extension Education, 28*(3, 4), 22–27.

Vijayaragavan, K., & Subramanyan, V. S. (1981). Information Source Credibility of Garden Land and Dry Land Farmers'. *Indian Journal of Extension Education, 17*(1, 2), 92–95.

CHAPTER 7

Agro-Tourism Potential in Kishtwar and Bhadarwah Region of Jammu and Kashmir

Madhu Bala

1 Introduction

For the overall economic transformation of any country, it is important to develop its agriculture sector because agro-tourism activities have significantly contributed to the rural economies in various parts of the world. For example, in Thailand millions of tourists visit farm areas; in Europe a large percentage of people take farm holidays; Greece has promoted mountainous and less favoured areas to attract tourists, and so on. (Bwana et al. 2015). Being a developing economy, India needs a strong agriculture sector to feed its growing population and meet the rising demands of the other sectors of the economy.

It is estimated that the tourism industry worldwide is growing at the rate of 4 per cent a year, whereas Indian tourism is growing at 10 per cent which is 2.5 times more than the world's rate (Ingavale 2015). India has a great potential to defeat the competition and increase the productivity of its tourism through agro-tourism. Agro-tourism is the fastest growing

M. Bala (✉)
Government Degree College Kishtwar, Kishtwar (J&K), India

H. Chahal et al. (eds.), *Sustainable Business Practices for Rural Development*, https://doi.org/10.1007/978-981-13-9298-6_7

sector in India, and states such as Karnataka, Kerala, Maharashtra, Jharkhand, Gujarat, Rajasthan, Himachal Pradesh and Goa have successfully implemented this concept, and it has been observed that the income level of the farmers and their standard of living have increased with new employment opportunities, and so on (Ubale and Borate 2012). It is a merger of two sectors, namely, agriculture and tourism, and provides an opportunity to the tourists to understand the rural life in India, to enjoy the food and to get to know the various farm activities during their visit (Ingavale 2015). Bansal et al. (2010) defined agro-tourism as 'a range of activities, services and amenities provided by farmers and rural people to attract tourists to their area in order to generate extra income for their business'.

The majority of our country lives in the rural areas, and their main sources of livelihood are agriculture, apiculture, horticulture, forestry, fishery and so on. According to Dr M.S. Swaminathan, major job opportunities could only be provided by agrarian activities (Barbuddhe and Singh 2014). Earlier, mountainous regions and beaches have been the centre of attraction, but in recent years, parks, bird sanctuaries and forests have attracted a number of tourists as well. Nowadays, people want to enjoy visiting rural life because such activities are affordable and less time-consuming (sometimes people can come back within a day), and people are also curious to see practically how and where things (such as fruits, vegetables, dairy, etc.) are produced. This provides an opportunity for the farmers to develop agro-tourism activities (Jagtap et al. 2010). This will not only address issues such as poverty, unemployment, regional inequality and so on (Mukherjee 2012) but will also benefit the farmer in raising their own income, sell 'experiences' of agricultural activities and also sell their products directly to the visitors. It would not only be advantageous to farmers but also to rural people and tourists as well. Some of the attractive features offered by the agro-tourism industry includes arts and crafts items (such as blankets made of sheep wool, carpet weaving and shoe making where both are made of either rice or wheat straw, embroidery, baskets made of straw, earthen pots, etc.), farm equipment, road-side selling of fresh farm products, display of various agricultural activities like sheep shearing, wool processing, fishing/hunting and so on; green meadows for picnics, various tours by educational institutions, herb walks, renting a tree, moonlight activities, lunch counter and so on, are also becoming centres of attraction for the tourists (Karri 2016).

Jammu and Kashmir, the northernmost state of India, is an important tourist destination across the world. It has an agro-climatic condition which makes it suitable for horticulture and floriculture (IBEF 2018). Each and every region of Jammu and Kashmir is unique in itself and one can experience diversity in its different regions—Kashmir and Jammu as a mountainous plain region and Ladakh as an extremely cold and dry terrain. Among these regions, Bhadarwah and Kishtwar (Jammu Division) are ideal locations for agro-tourism, as these areas are rich in natural resources and have the potential in various agro-cultural operations.

2 Objectives of the Study

- To promote agro-tourism in Bhadarwah and Kishtwar areas of Jammu and Kashmir.
- To observe the key areas hindering the development of agro-tourism.
- To make recommendations for the success of agro-tourism in this region.

3 Methodology

The research was mainly carried out through primary and secondary sources. The primary sources mainly included discussions with the field experts, and secondary source of data was collected from photographs, internet web, journals, books and so on. For the promotion of agro-tourism in Bhadarwah and Kishtwar, it was necessary to locate various tourist destinations which already existed and simultaneously examine the various types of agricultural activities prominent in these areas. Along with these agriculture-related activities, other activities such as traditional cultural activities, folk songs, dances, antique crafts, fairs, festivals and so on, were also identified. To find out various tourist locations in Bhadarwah and Kishtwar area, help of the field experts and internet were taken, and data related to agricultural pattern and various related activities were taken from the Government of Jammu and Kashmir; the Directorate of Economics and Statics; Planning Development and Monitoring Department, Kishtwar; and Planning Development and Monitoring Department, Doda.

4 Promotion of Agro-tourism in Bhadarwah and Kishtwar Area of Jammu and Kashmir

Barbuddhe and Singh (2014) suggested a systematic approach for the promotion of agro-tourism at any place. These included a survey of tourism sites, analysis of forest area and map identification of areas rich in flora and fauna, identification of plant biodiversity-rich location, analysis of agro-ecological areas, study on the time period, location of various fairs and cultural and religious festivals of the states. A number of sites have been identified by the tourist departments which are rich in flora and fauna. Some of them have been discussed further.

Potential Regions That Can Be Explored and Developed for Agri-Tourism in Kishtwar Region

Kishtwar is popularly known as the land of 'Sapphire and Saffron' since saffron flowers are cultivated here (it is considered to be the best produced saffron in India). Padder is known for sapphire deposits and other forest products like chilgoza (pine nuts), jeera (cumin seeds), guchhi (wild morchella), various herbs and so on. Tatta Pani, a hot spring, is also located here. Padder is visited by thousands of tourists especially during Machail Yatra. Handmade woollen blankets, shawl and gaba are made by the artisans here; silver bakerwal jewellery is also something this region is famous for. Annual fairs and festivals include urs, yatras and losar. Famous 'Gaddi' and 'Zagroo' dances are performed here. A national park (Kishtwar High Altitude National Park) is located on the northern side of Kishtwar district in Marwah subdivision. Marwah is also famous for trout fishing, mountaineering, springs and rock cut statue of a cow. Sarthal, 24 km from Kishtwar, is one of the most popular places frequently visited by tourists. Nagsen located east of Kishtwar town is full of natural beauty. This site is named after the Buddhist preacher Nagseen. Besides these, a fish hatchery at Mugalmaidan (chatroo) has been established. A total of 25,872 hectare area of the district is under food and non-food crops, out of which 15,186 hectare area is under high yielding variety seeds (HYVS) (Source: Government of Jammu and Kashmir 2015–2016). The area under cultivation of different crops, average horticultural produce and livestock/poultry population in Kishtwar district is given in Table 7.1. The following figures illustrate activities going on in some of the sites the author visited (Figs. 7.1 and 7.2).

Table 7.1 Table showing land use pattern under food crops and non-food crops, major horticulture crops (production in metric tonne) and livestock/poultry population

Sr. no.	*Crops*	*Area (in acres)*	*Horticulture crops*	*Production*	*Livestock/ poultry*	*Population (lakhs)*
1	Rice	1729	Apple	3029	Cattles	1.108
2	Maize	13,104	Pear	530	Buffaloes	0.138
3	Wheat	1958	Apricot	575	Sheep	2.289
4	Barley	1399	Peach	196	Goats	0.926
5	Grim/ millets	1268	Plum	185	Horse/ ponies	0.050
6	Pulses	1207	Grapes	7.85	Donkeys/ mules	0.069
7	Fodder crops	138	Walnut	6493	Yaks	0.155

Source: Government of Jammu and Kashmir (2015–2016)

Fig. 7.1 Saffron field at Kishtwar. (Source: Author)

Potential Regions That Can Be Explored and Developed for Agro-Tourism in Bhadarwah Region

Bhadarwah is designated as 'mini Kashmir' and is a mesmerising place in the Doda district of Jammu and Kashmir. It is blessed with vast grasslands and meadows, dense forest areas, diversified flora and fauna, beautiful valleys, streams and so on. Some of the attractive tourist spots include Chinta Valley, where horse riding is a popular activity for the tourists. Padri site

Fig. 7.2 Wheat field at Kishtwar. (Source: Author)

caters to many adventure activities such as skiing and paragliding, and thousands of pilgrims pass through this area to reach Mani-Mahesh temple. Trekking and horse riding are also done here. Jai is a perfect place for rock climbing enthusiasts and Jai nallah is popular for trout fish. A wide range of plants and animal species are found here. This valley is also famous for wild herbs such as Guchhi (wild Morchella), kasrode (fern) and so on, and Seoj meadow is also known for its majestic beauty. Bhal Padri is a beautiful valley comprising many small valleys and flowing rivulets and streams. Gatha is a tourist resort, known for the artificial lake, beautiful parks, tourist huts and traditional flour mill (Gharat). Bhadarwah is a blend of different cultures, art, language, fair and festivals and so on. 'Dheku' or 'Kudd' and 'Ghurai' folk dances are performed there and religious and historical songs are sung. The area under cultivation of different crops, average horticultural produce and livestock/poultry population in Bhadarwah is given in Table 7.2. The following figures illustrate activities going on in some of the sites the author visited (Figs. 7.3 and 7.4).

Karri (2016) suggested three vital principles of agro-tourism: first, that the visitors find something to see, something that will catch their eyes; second, that they can do or perform various activities themselves; and third, that they have something to buy as well. And all these three principles are applicable to Bhadarwah and Kishtwar region of Jammu and Kashmir. As there are tourist places to support agro-tourism, and tourists can see various kinds of animals, birds, herbs, festivals, folk dances, folk songs, traditional equipment and so on; they can participate in various agricultural activities such as fishing, harvesting, sowing, hunting, riding

Table 7.2 Table showing land use pattern under food crops and non-food crops, major horticulture crops (production in metric tonne) and livestock/poultry population

Sr. no.	*Crops*	*Area (in hectare)*	*Horticulture crops*	*Production*	*Livestock/ poultry*	*Population (lakhs)*
1	Paddy	442	Apple	801.00	Cattles	1.108
2	Maize	2270	Pear	298.00	Buffaloes	0.138
3	Pulses	182	Apricot	20.30	Sheep	2.289
4	Fodder	105	Peach	12.30	Goats	0.926
5	Vegetables	116	Plum	9.25	Horse/ponies	0.050

Source: Government of Jammu and Kashmir (2016–2017)

Fig. 7.3 Vast fields in Bhadarwah. (Source: D. K. Dogra 2018)

yak, horses and so on; and lastly, they can even buy rural crafts, raw honey, pulses, herbs and so on.

5 Key Areas Hindering the Development of Agro-tourism

- There is lack of agro-tourism awareness among the farmers
- Lack of skilled human capital

Fig. 7.4 Cattle grazing in the meadows. (Source: D. K. Dogra 2018)

- Inadequate rural infrastructures such as roads, pathways and so on
- Lack of training facilities for the farmers
- Social problems such as language, hygiene and so on
- Safety and security of the tourists (Barbuddhe and Singh 2014)

6 Recommendations for the Success of Agro-tourism

- There is a need to increase the awareness level of farmers towards agro-tourism and the opportunities it can offer in the form of additional income and employment, thereby benefitting the local people (Mukherjee 2012).
- Agro-tourism activities include tours of farms such as apple, grapes, apricot, pear and so on and bird and animal farms, poultry farms, dairy farms, goat farms and so on. For setting up such farms, a huge investment is required. The state government should take initiatives to support this sector through financial aid. For example, in Himachal

Pradesh, the government has launched a new initiative 'Har Gaon Ki Kahani' to explore the tourist potential of the villages (Parmar 2012).

- Basic infrastructure including roadways, airways and railways is essential for the success of agro-tourism because no one would like to visit a place which is difficult to reach.
- Establish a body responsible for the development, marketing and training of agro-tourism activities.
- A small-scale agro-tourism project could be initiated on an experimental basis.

7 Conclusion

It is evident from the preceding discussion that both Bhadarwah and Kishtwar have a distinctive geography and culture which can offer an opportunity to the state of Jammu and Kashmir in terms of growth and development of agro-tourism. Agro-tourism can act as a medium through which rural areas can be developed and provide an alternative source of income to the farmers and employment to the youth. When the villages will develop, India will develop.

References

Bansal, A., Thakur, S., & Mishra, S. (2010). Agricultural and Rural Tourism in India: Scopes in Current Environmental Changing Scenario. *Management Prudence Journal, 1*(1), 1–6.

Barbuddhe, S. B., & Singh, N. P. (2014). Agro-Eco Tourism: A New Dimension to Agriculture. *Technical Bulletin No. 46*, ICAR Research Complex For Goa, Old Goa, pp. 1–37.

Bwana, M. A., Olima, W. H. A., Andika, D., Agong, S. G., & Hayombe, P. (2015). Agritourism: Potential Socio-Economic Impacts in Kisumu Country. *IOSR Journal of Humanities and Social Science, 20*(3), 78–88.

Dogra, D. K. (2018). Photos. Retrieved from https://www.facebook.com/D.K.Dogra/photos

Government of Jammu and Kashmir. (2015–2016). *District Statistical Hand Book* (pp. 1–81). Kishtwar: Directorate of Economics and Statistics, Planning Development and Monitoring Department, District Kishtwar, District Statistics & Evaluation Office.

Government of Jammu and Kashmir. (2016–2017). *District Statistical Hand Book* (pp. 1–289). Doda: Planning Development and Monitoring Department, District Doda, District Statistics & Evaluation Office.

IBEF. (2018). About Jammu and Kashmir: Information on Tourism Industries, Economy & Geography. Retrieved from http://www.ibef.org/states/jammu-kashmir.aspx

Ingavale, D. (2015). Agri Tourism: A Business Model of Agri Tourism Development Corporation. *Paripex-Indian Journal of Research, 4*(7), 76–77.

Jagtap, M. D., Nichit, M. B., & Benke, S. R. (2010). Agri Tourism: The Performance, Problems and Prospects for the Farmers in Maharashtra. *International Journal of Commerce and Business Management, 3*(1), 153–156.

Karri, G. N. (2016). *Scope of Agritourism in India (with Reference to Development, Challenges, Extension & Advisory Services)* (pp. 1–40). Hyderabad: ICAR-National Academy of Agricultural Research Management.

Mukherjee, D. (2012). Rural Tourism Path to Economic & Regional Development in India. *Kurukshetra, 60*(7), 3–10.

Parmar, J. S. (2012). Tourism: An Engine to Economic Growth in the Rural Economy of Himachal Pradesh. *Kurukshetra, 60*(7), 7–20.

Ubale, N. B., & Borate, H. V. (2012). Agri Tourism: An Innovative Income Generation Avenue. *Kurukshetra, 60*(7), 21–25.

CHAPTER 8

Assessing the Factors Impacting Destination Loyalty in Sustainable Tourism: A Case Study of Sanasar Village, an Emerging Tourist Spot in Jammu and Kashmir (North India)

Neetu Kumari, Jhanvi Khanna, Pradeep Kumar, and Sandeep Patyal

1 Introduction

Tourism means temporarily travelling from usual environment, with the expectation of pleasure of novelty and change. The term tourist is used for those temporary visitors who stay in places different from their place of residence for more than one day. They mostly travel for the purpose of enjoying their vacations and seeing new places, for higher studies, medication,

N. Kumari (✉) • S. Patyal
University of Jammu, Jammu, Jammu and Kashmir, India

J. Khanna
Jammu and Kashmir State Board of School Education, Jammu, Jammu and Kashmir, India

P. Kumar
Government Degree College for Boys, Udhampur, India

H. Chahal et al. (eds.), *Sustainable Business Practices for Rural Development*, https://doi.org/10.1007/978-981-13-9298-6_8

visiting pilgrimages, meeting friends and family, business meetings, and so on (The United Nations Conference on International Travel and Tourism, 1963). The different ideologies and types of tourism followed all over India are sports tourism, health tourism, eco-tourism, medical tourism, sustainable tourism and accessible tourism. The Indian tourism sector is growing by leaps and bounds and India's contribution to world tourism sector is also increasing greatly, leading to rapid increase in tourist arrivals and foreign exchange earnings. The present study was undertaken in Sanasar, an emerging tourist spot in Udhampur, Jammu and Kashmir (J&K, India). In terms of opportunities, the area has the advantage of natural beauty, cheap labour, rich culture, affordable accommodation, transportation, communication and entertainment. In today's fast growing world, tourism can act as a catalyst for J&K to bring people from different parts of the country closer to each other. Quality, satisfaction, value and loyalty are the emerging concepts in tourism research (Rajesh 2013). Different researchers have different perceptions regarding sub-dimensions of these constructs (Christina Geng Qing Chi and Hailin Qu 2008). Therefore, there is a need to develop a model in order to measure these constructs in sustainable tourism and have a better understanding of the relationship between them (Oppermann 2000; Coban 2012).

2 Conceptual Framework

Tourism as a service industry comprises several allied activities; no single activity alone can perform this service. There are a number of useful dimensions that when taken together provides tourism service.

The first useful dimension is quality. In the tourism sector, quality signifies providing all the goods, services and facilities to the guests as per their expectations. Providing quality service is a pre-requisite for the growth and success of any hospitality market and delivering such a service is one of the biggest challenges faced by almost every country. The most important dimension that impacts a customer's perception about quality is 'accessibility'. The quality of reaching or entering any area safely and comfortably is accessibility. The term 'accessible' tourism implies an effort made to ensure that the tourist destination and various facilities are accessible to all, irrespective of any physical constraint. Important services and their accessibility affect the tourist's perception (Eusebio and Vieira 2011). The second dimension is 'physical environment quality'. The review of literature supports physical environment quality as a useful dimension of quality (Bitner 1992; Brady and Cronin 2001; Parasuraman et al. 1985;

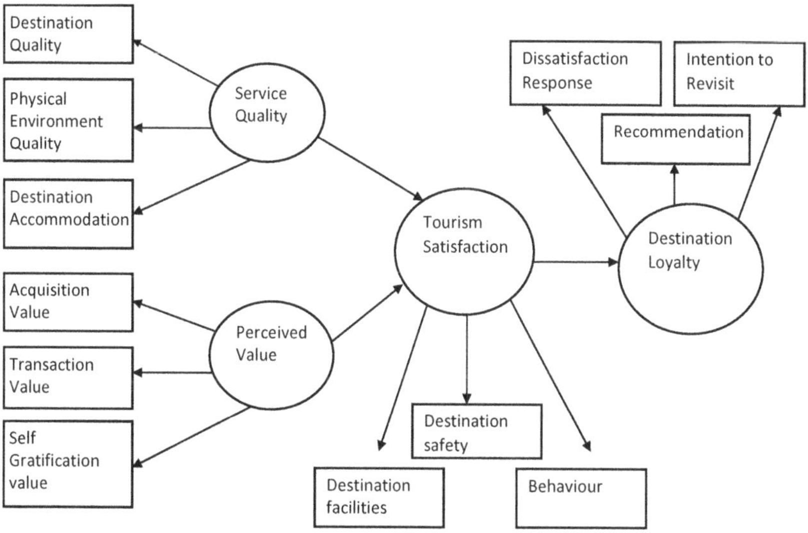

Fig. 8.1 Service quality, perceived value, tourist satisfaction, destination loyalty and sustainable tourism

Rust and Oliver 1994). The third sub dimension is 'destination accommodation', which comprises tangible equipment, physical facilities and the look of the accommodation (Conway and Willcocks 1997) where the tourist stays within tourist destination (See Fig. 8.1).

Hyp 1 : Perception about Destination Accessibility, Physical Environment Quality and Destination Accommodation positively influences the Service Quality.

According to Zeithaml, consumer perceived value (CPV) is a complete consumer evaluation of a good's utility depending upon its costs and benefits. It can be defined as consumer's complete evaluation of the usefulness of a product or service, depending upon the perception of what is given and what has been received. The first dimension of consumer perceived value is 'acquisition value'. Acquisition value means the perceived net gains that occur when products or services are taken and is known as a trade-off between utilities and disutilities (Chahal and Kumari 2012; Mathwick et al. 2002; Zeithaml 1988). The second dimension is 'transaction value', which acts as an important factor in defining the concept of perceived value

(Grewal et al. 1998; Parasuraman and Grewal 2000). Transaction value is the perception of psychological satisfaction achieved from a transaction or during the service delivery (Chahal and Kumari 2012). The last dimension is 'self-gratification value', which is an important component of CPV (Arnold and Reynolds 2003; Sweeney and Soutar 2001). Self-gratification value is the result of effective overall service transactions.

Hyp 2 : Acquisition value, Self-Gratification value and Transaction value positively influence the perceived value of the tourists.

Another important factor of the tourism sector is the satisfaction level of the tourists, which greatly affects the people's choice of destination (Kozak and Rimmington 2000). The first important dimension that affects tourist satisfaction is 'destination safety', which means how safe a tourist feels at the tourist destination. Further, 'destination facilities' attribute includes attractions such as lodging, dining, shopping, activities and events (Chi and Qing 2008). 'Attitude and behaviour' also affect customers (Brady and Cronin 2001; Chahal and Sharma 2004). Attitude as a superordinate concept of satisfaction was empirically demonstrated by Bitner (1992).

Hyp 3 : Destination Safety, Destination Facilities and Behaviour positively affect tourist satisfaction in the destination.

The intention of revisiting a place is the same as buying the same good repeatedly or purchasing the goods of the same brand (Tellis 1998). Loyalty can be defined as consumers' behaviour to repeatedly purchase similar goods or services (Hawkins et al. 1995). The first dimension is 'intention to revisit', where a person is given a whole lot of choices to buy the same good against all odds and at all costs (Jones and Taylor 2007). 'Recommendation' is another useful dimension, which means making other people aware about the positives of the place so that it motivates others also to visit that place. The cost of attracting new customers is lowered through just a word-of mouth from satisfied customers, which increases the company's reputation, or has an adverse effect due to dissatisfied customers (Anderson 1998; Fornell 1992). Lastly, 'dissatisfaction response' is important as it includes the various factors which act as a stumbling block or act as a barrier in revisiting a place due to various reasons which remained unsatisfied during their visit.

Hyp 4 : Intention to revisit, recommendation by other people and dissatisfaction response affects destination loyalty of the tourists.

Service quality draws the attention of researchers in the contemporary world and is attributed to its relationship with financial performance (Nelson et al. 1992), costs (Kellogg et al. 1997), customer satisfaction (Bolton and Drew 1991; Boulding et al. 1993; Cronin and Taylor 1992) and customer retention (Boshoff 1997). Customer satisfaction has become popular and an important construct because it is considered as a primary indicator of future profits and customer loyalty (Anderson and Fornell 2000; Bearden and Teel 1983; Oliver 1980). Oom et al. (2006) have studied the relationship between tourist satisfaction and destination loyalty. Customer loyalty has become an important research topic in contemporary era and an increase in customer loyalty and satisfaction secures future revenues and reduces the cost of future transactions (Bolton 1998; Fornell 1992; Reichheld and Sasser 1990; Rust and Oliver 1994). A satisfied customer is considered as an economic asset which has high return and low risk (Fornell et al. 2006). It has been seen that service quality and customer satisfaction, though theoretically different, are closely related constructs (Dabholkar et al. 1996).

Hyp 5 : Service quality and perceived value have direct positive effect on tourist satisfaction, which ultimately enhances the destination loyalty.

3 Research Methodology

Sample

The present study based on tourist loyalty considering perceived value, service quality and tourist satisfaction is based on primary data collected from tourists of Sanasar, a tourist spot in district Udhampur, Jammu and Kashmir, North India. Initially, pre-testing on 35 respondents was done for mainly two reasons. First, to check the content and face validity of the items for four different constructs. Second, it was carried out to find the sample size for the final data collection. The survey was conducted between December, January and February 2017–2018. The purposive sampling was used for data collection of 300 respondents.

Measures

All measures such as Service Quality, Perceived Value, Tourist Satisfaction and Destination Loyalty were extracted from reviewed literature of Brady and Cronin (2001), Chahal and Sharma (2004), Chahal and Kumari (2012). The vocabulary of scale items was tailored according to the tourism sector.

A five-point Likert-type scale was used for all the constructs, where (5) represented strongly agree and (1) represented strongly disagree.

Data Purification

For reducing the data, exploratory factor analysis was done, whereby some of the items were deleted from the initial set of items. Initially, response scores of negative items were changed and every item was checked for normal distribution. Then, from a statistical viewpoint, the item to total correlation coefficient was seen and those values were deleted, which were below other item–total correlations (0.03). It also helped in checking the role of inter-correlation among items, which is important for effective factor structure identification. For factor analysis, the study also used principal component analysis with varimax rotation as it is one of the best rotation procedures which maximises the number of items with high loadings on one factor, thus increasing the interpretability of the factors (Malhotra 2002). To determine the number of factors to be extracted, the Eigen value equal to or more than one criterion is used, and to find out relevancy of data reduction and grouping for factor analysis and Kaiser-Meyer-Olkin (KMO), value less than 0.50 is used. Further, Bartlett test of sphericity is used and degree of correlation coefficient less than 0.30 is used as criterion for selection of items (Hair et al. 2003). The pre-testing task leads to modification in the schedule by deleting eight items, which included four from service quality, three from tourist satisfaction and one item from destination loyalty. It was also checked that by deleting all these items, the intended meaning of the constructs of which they were a part was not harmed. Thus, finalised schedule comprised 15 items of SQ, 15 items of perceived value, 15 items of tourist satisfaction and 10 items of perceived loyalty.

Table 8.1 Sample and sub-sample wise Cronbach alpha values

	Overall	*Service quality*	*Perceived value*	*Tourist satisfaction*	*Destination loyalty*
Sample 300	0.960	0.855	0.945	0.850	0.900
Sub-sample I (1–150)	0.954	0.883	0.928	0.845	0.928
Sub-sample II (151–300)	0.940	0.800	0.900	0.844	0.935

Reliability and Validity

To know the extent to which the scale produces constant results of measurements, reliability of the scale is checked repeatedly (Hair et al. 2003) and to evaluate the reliability of the scale the internal consistency using split half method is used (Malhotra 2002; Tull and Hawkins 2005). The value of Cronbach alpha for the sample is found to be 0.960 and also the Cronbach alpha values for service quality, perceived value, tourist satisfaction and destination loyalty were 0.855, 0.945, 0.850, 0.900, respectively, showing the internal consistency and reliability of the sample. Also by dividing the respondents into two equal halves, the reliability of the data was examined. On the whole internal consistency was clearly visible amongst the responses of the respondents divided into two equal halves with alpha values around 0.8 (Table 8.1).

Validity: Validity of the scale items, including content, construct, convergent is examined at different stages.

Content validity: Through review of literature, discussions with the subject experts, tourists, native people, small and big businessmen, the content validity of the scale was evaluated at the time of pre-testing for the selection of items in the service quality and service performance constructs. Certain items were slightly changed to make them understandable to the respondents, which helped in checking the content and face validity.

Convergent validity: It assumes that measures of constructs are related to each other in the real sense, which should be related to each other theoretically as well. The significant correlation between sub-dimensions and their respective dimensions predicts a high degree of convergent validity (Table 8.2).

Table 8.2 Dimension-wise convergent validity of the scale

Dimensions	*Correlation value*	*Significance*
Service quality and destination accessibility	0.95	0.000
Service quality and physical environment quality	0.90	0.000
Service quality and destination accommodation	0.80	0.000
Perceived value and acquisition value	0.80	0.000
Perceived value and transaction value	0.85	0.000
Perceived value and self-gratification value	0.84	0.000
Tourist satisfaction and destination safety	0.92	0.000
Tourist satisfaction and destination facilities	0.90	0.000
Tourist satisfaction and behaviour	0.80	0.000
Destination loyalty and intention to revisit	0.80	0.000
Destination loyalty and recommendation	0.88	0.000
Destination loyalty and dissatisfaction response	0.82	0.000

Construct validity: To examine the construct validity of the scale, the communality values (greater than 0.5), Kaiser Meyer Olkin (KMO) Measure of Sampling Adequacy (MSA) value (greater than 0.7), variance explained (greater than 0.5) and factor loading values (greater than 0.5) are the criterion used (Hair et al. 2003). Most of the values met the threshold limit, which helped in checking the construct validity of the sub-scales (Table 8.4).

4 Demographic Profile

The sample of tourists in Sanasar (Udhampur, J&K) consisted of 90 females and 210 males. Majority of the tourists were young, whose age varied between 41 and 50 years (85). Majority of the tourists (220) were married. Further, majority of tourists were in service (110) and only 20 students were contacted. Income-wise, majority of respondents (120) belonged to income group 3 with income between Rs. 20,000 and 40,000 (See Table 8.3).

5 Exploratory Factor Analysis

The application of exploratory factor analysis resulted in the formulation of useful sub-dimensions for each measurement construct. The service quality contains KMO value of 0.85 and explains 60.55% variance. Service Quality is found to be a function of destination accessibility, physical

Table 8.3 Demographic profile of tourists

	Frequency	*% age*
Gender		
Female	90	30
Male	210	70
Age		
20–30 yrs (AG1)	70	23.33
31–40 yrs (AG2)	80	26.66
41–50 yrs (AG3)	85	28.33
51 and above	65	21.66
Marital Status		
Married	220	73.33
Unmarried	80	26.66
Occupation		
Service	110	36.66
Business	40	13.33
Unemployed	60	20
Student	20	6.66
Retired	60	20
Others	10	3.33
Income		
Below 20,000	40	13.33
20,000–40,000	120	40
40,000–80,000	80	26.66
80,000 above	60	20

environment quality and destination accommodation. The second dimension, perceived value, contains acceptable KMO (0.80) and explains 70.60% variance. The perceived value is found to be the function of acquisition value, transaction value and self-gratification value. The third dimension, tourist satisfaction (KMO = 0.65, variance explained = 65.05), is found to be the function of destination safety, destination facilities and behaviour. The last construct, destination loyalty, is a function of intention to revisit, recommendation and dissatisfaction response (Table 8.4).

6 Confirmatory Data Analysis

A confirmatory data analysis was applied after the exploratory factor analysis so as to examine the research measures and their reliability. Initially, for establishing unidimensionality of the sub-scale, indicators of measurement models of scale are examined, which is based on two criteria: regression weights (more than 0.50) and critical ratio (above 1.96) (Byrne 2001). To know the

Table 8.4 Factor-wise factor loading, KMO, MSA, communalities and variance explained for service quality, perceived value, tourist satisfaction and destination loyalty

Dimensions	*Mean values*	*Factor loading*	*KMO/ MSA*	*Variance %*	*Communalities values*
Service Quality			0.85	60.55	
F1 = Destination Accessibility				20.55	
Accessibility to tourist place	4.50	0.68	0.89		0.58
Connectivity to local tourist places	2.00	0.60	0.80		0.65
Proper Mobile phone networking	3.00	0.70	0.75		0.85
Money withdrawal facilities (e.g. ATM).	2.00	0.65	0.70		0.75
Proper parking facilities	4.00	0.70	0.79		0.50
F2 = Physical Environment Quality				18.20	
Ambience of tourist place	5.00	0.85	0.82		0.70
General cleanliness and sanitation	4.00	0.84	0.60		0.50
No Air, water, noise pollution	5.00	0.70	0.75		0.60
Availability of clean drinking water	2.00	0.72	0.82		0.50
Availability of toilets	3.00	0.60	0.70		0.65
Provision of benches/chairs for resting if needed	2.00	0.70	0.50		0.70
F3 = Destination Accommodation				21.80	
Quality of the accommodation	4.00	0.80	0.60		0.50
Ambience of the spot and surroundings of the accommodation	4.50	0.70	0.65		0.65
Ambience and cleanliness of the inside of the accommodation	4.00	0.75	0.70		0.70
Facilities/equipment of the accommodation	3.50	0.60	0.68		0.55
Perceived Value			0.80	70.60	
F1 = Acquisition value				30.05	
Value for money	3.60	0.70	0.65		0.55
Value for money of restaurants, cafés and bars you visited	3.80	0.65	0.60		0.58
Affordable cost of activities	3.90	0.70	0.70		0.45

(*continued*)

Table 8.4 (continued)

Dimensions	*Mean values*	*Factor loading*	*KMO/ MSA*	*Variance %*	*Communalities values*
Deliver value-added services at reasonable cost	3.89	0.65	0.72		0.65
Less walking and excursions	4.00	0.68	0.65		0.60
F2 = Transaction value				22.46	
Price of general commodities in the tourist place	2.00	0.80	0.65		0.60
Hotels have no hidden charges for their services	3.00	0.78	0.70		0.65
Price of Local Transport	3.50	0.75	0.60		0.72
Discount in tour packages	3.60	0.65	0.68		0.70
No begging and cheating in the tourist place	4.40	0.70	0.72		0.68
Reasonable services	4.00	0.72	0.70		0.70
Money worth in this trip	4.50	0.74	0.72		0.68
F3 = Self-gratification value				18.09	
You got relief from stress by visiting this place	4.50	0.80	0.78		0.70
You got psychological satisfaction	4.00	0.78	0.75		0.70
You got a chance to socialise	4.80	0.75	0.72		0.68
Tourist Satisfaction			0.65	65.05	
F1 = Destination safety				27.16	
You are satisfied with Law and Order of place	4.00	0.70	0.68		0.68
Safety of luggage	4.00	0.72	0.65		0.70
Safety of females	3.48	0.70	0.60		0.63
F2 = Destination facilities				20.09	
Satisfied with transport facility	2.00	0.70	0.72		0.60
Satisfied with shopping facility	3.00	0.72	0.70		0.70
Satisfied with sanitation facility	3.60	0.65	0.68		0.65
Healthcare and medical services on tourist place	2.00	0.60	0.65		0.58
Behaviour and professionalism of the staff and owners in accommodation	3.00	0.72	0.60		0.65
Games and Activities—golf, gym, sports, casino etc.	2.00	0.70	0.65		0.60
Availability of Tourism Information Centre	2.00	0.68	0.65		0.62

(*continued*)

Table 8.4 (continued)

Dimensions	*Mean values*	*Factor loading*	*KMO/ MSA*	*Variance %*	*Communalities values*
Satisfied from services of tour guide	3.00	0.70	0.68		0.65
F3 = Behaviour				17.80	
Behaviour of other tourists	4.20	0.80	0.75		0.70
Behaviour of local people	4.00	0.75	0.70		0.60
Behaviour of taxi/auto-rickshaw drivers	4.45	0.70	0.70		0.65
Customer service of restaurants, cafés and bars you visited	4.00	0.68	0.65		0.60
Destination Loyalty			0.80	50.80	
F1 = Intention to revisit				21.14	
You will visit this place again	4.45	0.80	0.75		0.70
You will come again to other tourist places here	4.00	0.89	0.72		0.65
You are loyal to this tourist place	4.30	0.85	0.80		0.60
F2 = Recommendation				15.60	
Recommend this destination to your friends and relatives	4.00	0.80	0.75		0.60
Recommend this to your Internet friends	4.00	0.78	0.70		0.63
Recommend hotel to your friends and relatives	4.20	0.72	0.68		0.62
F3 = Dissatisfaction response				14.06	
In case of any problem complain to competent government authorities	4.00	0.70	0.70		0.60
In case of any problem discuss with other tourists	4.00	0.72	0.72		0.70
Complaint feedback to external agencies	3.00	0.65	0.68		0.70
In case of any problem complain to hotel authorities	4.54	0.69	0.65		0.50

suitability of the confirmatory factor analysis, composite reliability (greater than 0.7) and average variance extracted (greater than 0.5) of the sub-scales are assessed. This is followed by assessment of the model fit indices.

For testing the model, the criteria used are RMSEA within the acceptable range of 0.08 or less, Comparative Fit Index (CFI) and Non-formal Fit Index (NFI) values greater than 0.09, Chi square/df value less than 5,

Table 8.5 Composite reliability and average variance extracted of scale

Dimensions	*Composite reliability*	*Average variance extracted*
Service quality	0.90	0.90
Perceived value	0.85	0.80
Tourist satisfaction	0.75	0.82
Destination loyalty	0.80	0.88

and standardised regression weights and critical ratios above the threshold values.

7 Measurement Model

Unidimensionality: The critical value of the sub-dimensions of all the indicators, that is, service quality, perceived value, tourist satisfaction and destination loyalty are significant (greater than 1.96). Also the average variance extracted was greater than 0.5, showing convergent validity of the scale and unidimensionality of the sub-scales. The average variance extracted for the latent variables was more than 0.8, which shows high level of construct reliability.

Convergent validity: To establish convergent validity, the recommended average variance extracted value is 0.50 or more. The Average Variance Extracted (AVE) of service quality (0.90), perceived value (0.80), tourist satisfaction (0.82) and destination loyalty (0.88) indicate convergent validity of the scale (Table 8.5).

Composite reliability: Using Structural Equational Modelling, the composite reliability of the sub-scales is assessed (Hair et al. 2003). All the values, that is, 0.90 for service quality, 0.85 for perceived value, 0.75 for tourist satisfaction and 0.80 for destination loyalty (Table 8.5), are more than the threshold criteria.

8 Structural Model

After developing the scale for measuring service quality, perceived value, tourist satisfaction and destination loyalty, the study analyses the relationship between dimensions and their sub-dimensions and its overall model.

The critical ratio values above 1.96 at 95% level of significance reveal that destination accessibility, physical environment quality and destination accommodation significantly affect service quality. Hence, Hyp 1 is accepted. Further, acquisition value, transaction value, self-gratification value positively affect perceived value having acceptable value and hence Hyp 2 is accepted. Further, destination safety (SRW = 0.78), destination facilities (SRW = 0.70), and behaviour (SRW = 0.75) have acceptable CR values and SRW; hence, they showed positive and significant effect on tourist satisfaction. Lastly, values above 14.05 CR and 0.7 SRW showed that there is positive and significant effect of intention to revisit, recommendation, and dissatisfaction response on destination loyalty. Hence Hyp 4 is accepted.

Effect of service quality and perceived value on tourist satisfaction and destination loyalty is significant, as shown in Fig. 8.1. The relationship of the respective factors with the components is strong and the regression loadings fall between 0.7 and 0.8. The results indicate positive relationship between the dimensions. The model fit indices show CMN/df (25.06), RMSEA (0.060), NFI (0.80), CFI (0.82) (Table 8.6). The results depict good model fit. Hence, Hyp 5 is accepted.

Table 8.6 CR, RW, R2 of service quality, perceived value, tourist satisfaction and destination loyalty

Components	*Dimensions*	*CR*	*SRW*	*R2*
Service quality	Destination quality	20.90	0.80	0.65
	Physical environment quality	Ref	0.80	0.60
	Destination accommodation	17.08	0.75	0.62
Perceived value	Acquisition value	25.50	0.78	0.70
	Transaction value	Ref	0.70	0.65
	Self-gratification value	25.08	0.80	0.75
Tourist satisfaction	Destination safety	20.09	0.78	0.72
	Destination facilities	18.00	0.70	0.70
	Behaviour	12.08	0.75	0.70
Destination loyalty	Intention to revisit	19.15	0.74	0.72
	Recommendation	30.01	0.80	0.78
	Dissatisfaction response	14.05	0.70	0.69
Model fit	Chi square Equivalent in Confirmatory Factor Analysis (CMIN/DF) = 25.06	RMSEA = 0.060	Non-formal Fit Index (NFI) = 0.80	Comparative Fit Index (CFI) = 0.82

9 Discussion

The four-factor model given by the study is an important contribution for both theory and practice. The findings of the study show that destination loyalty, perceived value, service quality and tourist satisfaction are critical factors that affect tourist perception.

The three dimensions, destination accessibility, followed by physical environment quality and destination accommodation, contribute towards service quality. The results indicate that accessibility to tourist place, connectivity, proper mobile phone network, money withdrawal facilities, parking facilities contribute towards destination accessibility. The tourists are also conscious about physical environment quality, which is the function of ambience of tourist place, general cleanliness and availability of clean drinking water, toilets and provision of benches/chairs for resting. The second important dimension that impacts tourists' perceived value is the function of acquisition value, which is, value for money, affordable cost of activities, value-added services. The second dimension, transaction value, is the function of price of general commodities with no hidden charges, availability of local transport, discounts, absence of begging and cheating, reasonable services, and money worth in trip. Lastly, self-gratification value is a function of relief from stress by visiting a place, psychological satisfaction and socialisation. Further, tourist satisfaction is an important dimension which can help in gaining tourist loyalty. Tourist satisfaction comprises destination safety, which includes safety of family and luggage. The second dimension is destination facilities, which is a function of transport facility, shopping facilities, sanitation facilities, healthcare services, games and activities, tourism information centre and tourist guide. Lastly, behaviour is an important dimension that impacts tourist satisfaction, which includes behaviour of other tourists, local people, taxi/auto-rickshaw drivers, customer services. Destination loyalty is the last and most important factor used in the study, which is the function of intention to revisit, recommend and lodge complaints in case of dissatisfaction. The intention to revisit includes visiting a place again and the tourist's loyalty to a specific tourist place. Recommendation is the second dimension, which is a function of recommending a place to your friends, relatives and virtual friends. Lastly, dissatisfaction response is when the tourist faces a problem and he/she lodges a complaint with the competent government authorities, hotel authorities, external agencies and discusses with other tourists. The findings of the study support the idea that perceived value and service quality result in tourists' satisfaction and destination loyalty.

10 Managerial Implications

From the marketing point of view, the marketer must design tourism product strategy by considering the following marketing characteristics of tourism services. With respect to service quality, tourists are specifically dissatisfied with connectivity to local tourist places (M.S. = 2.00); especially during winters, traffic jam is the most common problem faced by tourists, so it is essential to widen the roads. Second, money withdrawal facilities (M.S. = 2.00) also impact negatively; so more ATMs can be placed with the help of private and public banks. In physical environment quality, tourists are not getting drinking water facilities (M.S = 2.00), so water facilities should be provided at the tourist destination. The tourists also complained that there is no provision of benches/chairs for resting if needed. Further, tourist satisfaction is also impacted by unavailability of healthcare and medical facilities (M.S. = 2.00), games and activities (M.S. = 2.00), availability of tourism information centre (M.S. = 2.00), which should be considered by tourism authorities. Further, the tourists strongly show intention to revisit, recommend and there is positive word of mouth for the tourist destination. In addition, the perceived value of the tourist perception is quite positive, which shows that Sanasar is an emerging tourist spot which has all the capabilities to be a national tourist spot in India.

11 Study Limitations and Future Research

Although a lot of effort went into this study, there were some limitations. First, the study is limited to only one tourist spot located in Udhampur district of Jammu and Kashmir (North India), which means that the study is restricted to a specific geographical area and also the data was analysed only through 200 questionnaires. This study may be the first attempt to assess the relative impact and inter-relationship of the four constructs and their sub-dimensions in tourism. It included only a global measure and future studies exploring this issue should include more measures. For further studies, researchers can add destination image as an important dimension and can examine its relationship with other constructs. Also, the study was based on convenience sampling method; therefore, its findings may not be representative of the entire population. Thus, to generalise the results of this study, an attempt should be made to include more tourist destinations in different parts of the country.

Appendix

Questionnaire

1) Nationality:

2) Gender:

3) Age:

4) Marital Status: Married ( ) Unmarried ( )

5) Occupation: Service ( ) Business ( ) unemployed ( ) Student ( ) Retired ( ) others, then specify

6) Year of visiting ....

7) Purpose of Visit (Tick the suitable column)

Leisure /recreation holidays ☐

Visiting friends and relatives ☐

Business ☐

Others (Please specify).............

8) How did you get information about this destination? (More than one answer allowed)

Travel agency ☐

Internet Family/friends ☐

Newspapers/magazines ☐

Travel brochures ☐

Travel guides ☐

TV ☐

Radio ☐

Tourist fairs ☐

Other: ....

9) How many times have you visited this place?(One/ Two/Three/ Four or more times).

10) Are you likely to come again and again to this tourist place? (Yes/ No)

11) Which season is the best for visiting this place? (Winter / Summer/ Rainy)

12) Do you face any language problem? (Yes/ No) Your mother tongue---------------

	Items
	Service Quality
	Destination accessibility
Q1	Accessibility to tourist place
Q2	Connectivity to local tourist places
Q3	Proper mobile phone networking
Q4	Money withdrawal facilities (e.g. ATM).
Q5	Proper parking facilities
	Physical environment quality
Q6	Ambience of tourist place
Q7	General cleanliness and sanitation
Q8	Air, water, noise pollution
Q9	Availability of clean drinking water
Q10	Availability of toilets
Q11	Provision of benches/chairs for resting if needed
	Destination accommodation
Q12	Quality of the accommodation
Q13	Ambience of the spot and surroundings of the accommodation
Q14	Ambience and cleanliness of the inside of the accommodation
Q15	Facilities/equipment of the accommodation
	Perceived Value
	Acquisition value
VA1	Value for money
VA2	Value for money of restaurants, cafés and bars you visited
VA3	Affordable cost of activities
VA4	Deliver value-added services at reasonable cost
VA5	Walking and excursions
	Transaction value
VT6	Price of general commodities in the tourist place
VT7	Hotels will have no hidden charges for their services
VT8	Price of local transport
VT9	Discount in tour packages
VT10	Begging and cheating in the tourist place
VT11	Reasonable services
VT12	Money worth in this trip
	Self-gratification value
VSG13	You got relief from stress by visiting this place
VSG14	You got psychological satisfaction
VSG15	You got opportunity to socialise
	Tourist Satisfaction
	Destination safety
SS1	You are satisfied with law and order of place
SS2	Safety of luggage
SS3	Safety of females

(*continued*)

(continued)

	Items
	Destination facilities
SF4	Satisfied with transport facility
SF5	Satisfied from shopping facility
SF6	Satisfied from sanitation facility
SF7	***Healthcare and medical services in tourist place
SF8	Behavior and professionalism of the staff and owners in accommodation
SF9	Games & Activities—golf, gym, sports, casino etc
SF10	Availability of Tourism Information Center
SF11	Satisfied from services of Tour guide
	Behaviour
SB12	Behaviour of other tourists
SB13	Behaviour of local people
SB14	Behaviour of taxi/Auto-rickshaw drivers
S15	Customer service of restaurants, cafés and bars you visited
	Destination Loyalty
	Intention to revisit
LV1	You will visit this place again
LV2	You will come again to other tourist places here
LV3	You are loyal to this tourist place
	Recommendation
LR4	Recommend this destination to your friends and relatives
LR5	Recommend this to your Internet friends
LR6	Recommend hotel to your friends and relatives
	Dissatisfaction response
LD7	In case of any problem complaint to competent government authorities
LD8	In case of any problem discuss with other tourists
LD9	Complaint feedback to external agencies
LD10	In case of any problem complaint to hotel authorities

References

Anderson, E. W. (1998). Customer Satisfaction and Word of Mouth. *Journal of Service Research, 1*, 5–17.

Anderson, E. W., & Fornell, C. (2000). Foundations of the American Customer Satisfaction Index. *Total Quality Management, 11*(6), 869–882.

Arnold, M. J., & Reynolds, K. E. (2003). Hedonic Shopping Motivations. *Journal of Retailing, 79*, 77–95.

Bearden, W. O., & Teel, J. E. (1983). Selected Determinants of Consumer Satisfaction and Complaint Reports. *Journal of Marketing Research, 20*(1), 21–28.

Bitner, M. J. (1992). Evaluating Service Encounters: The Effects of Physical Surrounding and Employee Responses. *Journal of Marketing, 54*(2), 69–81.

Bolton, R. N. (1998). A Dynamic Model of the Duration of the Customer's Relationship with a Continuous Service Provider: The Role of Customer Satisfaction. *Marketing Science, 17*, 45–65.

Bolton, R. N., & Drew, J. H. (1991). A Longitudinal Analysis of the Impact of Service Changes on Customer Attitudes. *Journal of Marketing, 55*, 1–19.

Boshoff, C. (1997). An Experimental Study of Service Recovery Options. *International Journal of Service Industry Management, 8*(2), 110–130.

Boulding, W., Kalrs, A., Staelin, R., & Zeithaml, V. A. (1993). A Dynamic Process Model of Service Quality: From Expectations to Behavioral Intentions. *Journal of Marketing Research, 30*(1), 7–27.

Brady, M. K., & Cronin, J. J. (2001). Some New Thoughts on Conceptualizing Perceived Service Quality: A Hierarchical Approach. *Journal of Marketing, 65*, 34–49.

Byrne, B. M. (2001). *Structural Equation Modelling with AMOS Basic Concepts, Applications and Programming* (pp. 78–87). New Jersey: Lawrence Erlbaum.

Chahal, H., & Kumari, N. (2012). Service Quality and Performance in the Public Health Care Sector. *Health Marketing Quarterly, 29*(3), 181–205.

Chahal, H., & Sharma, R. D. (2004). Managing Health Care Service Quality in a Primary Health Care Sector. *Metamorphosis, 3*(2), 112–131.

Chi, C. G., & Qing, Q. H. (2008). Examining the Structural Relationships of Destination Image, Tourist Satisfaction and Destination Loyalty: An Integrated Approach. *Journal of Tourism Management, 29*, 624–636.

Chi, C. G. Q., & Qu, H. (2008). Examining the Relationship Between Tourist's Attribute Satisfaction and Overall Satisfaction. *Journal of Hospitality Marketing and Management, 18*(1), 4–25.

Coban, S. (2012). The Effects of the Image of Destination on Tourist Satisfaction and Loyalty: The Case of Cappadocia. *European Journal of Social Science, 29*(2), 222–232.

Conway, T., & Willcocks, S. (1997). The Role of Expectations in the Perception of Health Care Quality: Developing a Conceptual Model. *International Journal of Health Care Quality Assurance, 10*(3), 131–140.

Cronin, J. J., & Taylor, S. A. (1992). Measuring Service Quality: A Re-examination and Extension. *Journal of Marketing, 56*(3), 55–69.

Dabholkar, P. A., Thorpe, D. I., & Rentz, J. O. (1996). A Measurement of Service Quality for Retail Stores Development and Validation. *Journal of Academy of Marketing Science, 24*(1), 3–16.

Eusebio, C., & Vieira, A. L. (2011). Destination Attributes' Evaluation, Satisfaction and Behavioural Intentions: A Structural Modelling Approach. *International Journal of Tourism Research, 15*(1), 66–80.

Fornell, C. (1992). A National Customer Satisfaction Barometer: The Swedish Experience. *Journal of Marketing, 56*, 6–21.

Fornell, C., Mithas, S., Morgeson, F. V., & Krishnan, M. S. (2006). Customer Satisfaction and Stock Prices: High Returns, Low Risk. *Journal of Marketing, 70*(1), 3–14.

Grewal, D., Manroe, K. B., & Krishan, R. (1998). The Effects of Price-Comparison Advertising on Buyer's Perceptions of Acquisition Value, Transaction Value and Behavioral Intentions. *Journal of Marketing, 62*, 46–59.

Hair, J. F., Anderson, R. E., Tatham, R. L., & Black, W. C. (2003). *Multivariate Data Analysis.* New Delhi: Pearson Education.

Hawkins, D. I., Best, R. J., & Coney, K. A. (1995). *Consumer Behaviour: Implications for Marketing Strategy* (6th ed.). Chicago: Irwin.

Jones, T., & Taylor, S. F. (2007). The Conceptual Domain of Service Loyalty: How Many Dimensions. *Journal of Service Marketing, 21*(1), 36–51.

Kellogg, D. L., Youngdahl, W. E., & Bowen, D. E. (1997). On the Relationship between Customer Participation and Satisfaction: Two Frameworks. *International Journal of Service Industry Management, 8*(3), 206–219.

Kozak, M., & Rimmington, M. (2000). Tourist Satisfaction with Mallorca, Spain, as an Off-season Holiday Destination. *Journal of Travel Research, 38*(3), 260–269.

Malhotra, N. K. (2002). *Marketing Research: An Applied Orientation.* New Delhi: Pearson Education.

Mathwick, C., Malhotra, N. K., & Rigdon, E. (2002). The Effect of Dynamic Retail Experiences on Experiential Perceptions of Value: An Internet and Catalog Comparison. *Journal of Retailing, 78*, 51–60.

Nelson, E., Rust, R. T., Zahorik, A. J., Rose, R. L., & Seimanski, B. A. (1992). Do Patients Perception of Quality Relate to Hospital Financial Performance? *Journal of Health Care Marketing, 3*(4), 1–13.

Oliver, R. L. (1980). A Cognitive Model of the Antecedents and Consequences of Satisfaction Decisions. *Journal of Marketing Research, 17*, 460–490.

Oom, P., Silva, J. A., Mendes, J., & Guerreiro, M. (2006). Tourist Satisfaction and Destination Loyalty Intention: A Structural and Categorical Analysis. *International Journal of Business Science and Applied Management, 1*(1), 26–44.

Oppermann, M. (2000). Tourism Destination Loyalty. *Journal of Travel Research, 39*, 78–84.

Parasuraman, A., & Grewal, D. (2000). The Impact of Technology on the Quality-Value-Loyalty Chain: A Research Agenda. *Journal of the Academy of Marketing Science, 28*(1), 168–174.

Parasuraman, A., Zeithaml, V. A., & Berry, L. L. (1985). A Conceptual Model of Service Quality and its Implication for Future Research. *Journal of Marketing, 49*(4), 41–50.

Rajesh, R. (2013). Impact of Tourist Perceptions, Destination Image and Tourist Satisfaction on Destination Loyalty: A Conceptual Model. *PASOS, 11*(3), 67–78.

Reichheld, F. F., & Sasser, W. E. (1990). Zero Defections: Quality Comes to Services. *Harvard Business Review, 68*, 105–111.
Rust, R. T., & Oliver, R. L. (1994). *Service Quality: New Directions in Theory and Practice* (pp. 1–19). Newbury Park, CA: Sage.
Sweeney, J. C., & Soutar, G. N. (2001). Consumer Perceived Value: The Development of a Multiple Item Scale. *Journal of Retailing, 77*(2), 203–220.
Tellis, G. J. (1998). Advertising Exposure, Loyalty and Brand Purchase: A Two Stage Model of Choice. *Journal of Marketing Research, 25*, 134–144.
Tull, D. S., & Hawkins, D. I. (2005). *Marketing Research-Measurement and Methods* (6th ed.). New Delhi: Printing Hall of India.
Zeithaml, V. A. (1988). Consumer Perceptions of Price, Quality and Value: A Means-end Model and Synthesis of Evidence. *Journal of Marketing, 52*, 2–22.

CHAPTER 9

How Leader's Proclivity and Internal Communication Affect Rural Employees' Attitude?

Hardeep Chahal and Saguna Sethi

1 Introduction

In today's competitive environment, it is vital for organisations to keep pace with the rapidly changing environment to sustain competition. Besides the teaching faculty, strong, capable and innovative supportive staff members are considered as important assets for achieving sustainability and objectives. Moreover, behaviour of higher authorities (i.e. leaders) also affects supportive staff/employees' ability to perform organisational functions effectively and efficiently. Lack of positive reinforcement from the leaders puts a question mark in front of employees who are forced to think whether to place their loyalties with the organisation or quit working there. For instance, if leaders show a negative attitude and block internal communication channels in the organisation, subordinates' mindset and attitude also become negative towards their leaders and the job, thereby

H. Chahal • S. Sethi (✉)
Department of Commerce, University of Jammu, Jammu,
Jammu and Kashmir, India

H. Chahal et al. (eds.), *Sustainable Business Practices for Rural Development*, https://doi.org/10.1007/978-981-13-9298-6_9

reducing their level of commitment. Employees' dissatisfaction from job ultimately causes organisations to suffer in the long term.

In managerial context, leaders' proclivity reflects the ways in which management approaches various organisational and employees' issues and problems. Srivastava and Frankwick (2011) stated that management's attitude is reflected through how effectively leaders promote organisational learning and how they handle contingent situation due to environmental instability. It is also a well-known fact that leader's negative, reactive, unjust and attitudes bring bad consequences not only for concerned group of employees but for the entire organisation (Pinder and Harlos 2001). In addition to leaders' proclivity, internal communication quality in the organisation is equally important to create an amiable environment for the growth and development of the organisation in general and for the supportive staff in particular. The effectiveness of internal communication between the leaders and the subordinates in the organisation can be evaluated by identifying whether the existing wide range of practices can make the mutual sharing of concerns, information and knowledge among employees possible (Hatipoglu and Inelmen 2017). Organisations that support participatory climate encourage internal communication by both formal as well as informal voicing platforms (Huang et al. 2005). Formalised practices such as grievance processes, self-managed teams and empowerment by the supervisor and informal practices including face-to-face discussions, one-on-one meetings and open-door policies support the smooth flow of internal communication within the organisation (Hatipoglu and Inelmen 2017). The existing literature supports that internal communication in the form of formal voice mechanisms, communication opportunities and freedom to form informal networks significantly influences employees' attitude and behaviour (Dedahanov et al. 2016; Van Dyne et al. 2003; Willman et al. 2006). Favourable outcomes are observed for employees if they are provided effective platforms to communicate within the organisation (Jones and Kelly 2014). On the other hand, the root cause of failure of several organisations is unsupportive, irresponsive management; inadequate and ineffective internal communication channels in the organisation (Jones and Kelly 2014; Allard-Poesi and Hollet-Haudebert 2017; An and Bramble 2017). Employees in such cases assume that management does not care about them and leaders are not interested in hearing any suggestions or criticisms and prohibit any negative feedback (Jones and Kelly 2014). Moreover, Morrison and Milliken (2000) and

Vakola and Bouradas (2005) observed that at some workplaces top management is also reluctant to offer a helping hand and is not ready to share information with subordinates, thereby, encouraging unfavourable management practices. This creates a dead effect in the entire organisation by losing employees' satisfaction and commitment towards their job forever.

In brief, urban workplaces also comprise a diverse workforce that comes from a rural background to work and with the hope of getting job satisfaction. To secure commitment from them, rural employees' valid expectations and desires must be fulfilled by the management from time to time. Many researchers suggested the need for research in managerial characteristics such as leader's attitude, position, status, styles, and so on, that affect employees' attitude towards their job (Guenter et al. 2017; Kish-Gephart et al. 2009; Park et al. 2016; Vakola and Bouradas 2005; Xu et al. 2015). Moreover, researchers such as Venkataramani et al. (2016) and Hatipoglu and Inelmen (2017) suggested that we need to study the effectiveness of mechanisms for internal communication inside the organisation. With this central idea, this chapter provides an insight into the relationship between leader's attitude and rural employees' job attitude. Therefore, the specific objective of the study includes the examination of the impact of leader's attitude and internal communication on employees' job satisfaction and commitment (see Fig. 9.1). Moreover, the study contributes to the literature by examining the relationships (in the emerging context of India) for rural employees in the higher education sector.

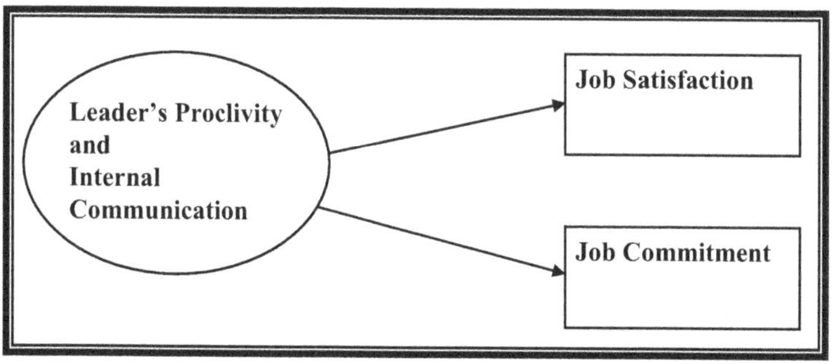

Fig. 9.1 The conceptual framework. (Source: Authors' research model)

2 Literature Review and Hypotheses Development

Leader's Proclivity and Employees' Job Attitude

Proclivity plays a very important role in determining the way in which a person responds, either positively or negatively to his or her environment. A leader's friendly behaviour or proclivity/attitude provides encouragement to employees, whereas, a leader's avoidance attitude negatively affects employee behaviour (Venkataramani et al. 2016). Furthermore, a leader's negative attitude towards subordinates can be observed in many ways such as in their beliefs about subordinates, that is, them being untrustworthy, selfish, and so on (Morrison and Milliken 2000). It is also observed that sometimes management deliberately reacts negatively by becoming unresponsive to the concerns of subordinates due to cost–benefit analysis (Willman et al. 2006). The existing literature recognises that managerial factors such as top management and supervisor's leadership and their attitude significantly affect employee job attitude. In this context, Yousef (2000) has also confirmed that leadership behaviour significantly affects satisfaction and job commitment. In addition, Van Vuuren et al. (2007) explained that management's positive attitude, which is reflected in the form of taking feedback from and listening to what employees want to convey, increases the level of satisfaction and commitment among them. Further, Thomas et al. (2006) proved that organisations where management positively responds to employees' need such as communication, learning and work schedule flexibility has a positive effect on organisational commitment. Moreover, Ruppel and Harrington (2000) confirmed that fair, just and right kind of treatment provided by the management secures employees' trust and commitment by developing their positive perception about the organisational unit. Very recently, Kim and Beehr (2018) also examined that leaders who adopt positive behaviour or attitude towards their subordinates by providing empowerment to them secure affective organisational commitment. Further, Orpen (1997) found that leaders that involve employees in the important matters concerning their job and provide motivation to them by taking care of their communication needs are more satisfied in their jobs. Similarly, Babin and Boles (1996) also confirmed that supervisor's attitude promoting a supportive environment and a good co-worker involvement helps in reducing employees' stress and increasing their job satisfaction. The existing literature suggests that some leaders adopt negative leadership attitude towards their followers such as Machiavellianism (Erkutlu and Chafra 2019) whereas some leaders adopt positive leadership styles. In the context of positive attitude of leaders' towards their follower, Chen et al. (2005) examined perceptions of nursing deans' and directors' leadership style in Taiwan and found that leadership style significantly affects job satisfaction. Further, Madlock (2008) confirmed supervisor's leadership style is significantly and positively related with employee job satisfaction. Further, Asrar-ul-Haq and Kuchinke (2016) observed positive relationship between transformational leadership and employee satisfaction in their study of the banking sector of Pakistan. Moreover, Ampofo (2016) revealed significant and positive association between transformational leadership style and various forms of commitment (affective, continuance and normative). Therefore, from the previous discussion on leader's proclivity that is reflected in the form of leadership attitude, style, behaviour, responsiveness and various beliefs regarding subordinates, it is hypothesised that:

Hypothesis 1 : Leader's proclivity positively affects employees' job attitude (which consists of satisfaction and commitment).

Internal Communication and Employees' Job Attitude

When employees perceive that supervisors at various positions possess interpersonal communication skills and make use of mechanisms to communicate, it increases their positive job attitude such as commitment towards the organisation (Thomas et al. 2006; Bambacas and Patrickson 2008; Ruppel and Harrington 2000). Many researchers such as Orpen (1997), Van Vuuren et al. (2007), Madlock (2008), Carriere and Bourque (2009), Men (2014) and Jacobs et al. (2016) also confirmed the positive association between internal communication and employees' job satisfaction. Furthermore, Jacobs et al. (2016) found that internal communication significantly and positively influences internal integration, external integration and satisfaction among employees. Moreover, some researchers also found that internal communication is positively related to both employee satisfaction and commitment. Nobile (2017) confirmed in primary school in Western Australia that communication opportunities provided to the staff members positively influenced their level of job satisfaction and job commitment. From the earlier discussion it can be hypothesised that:

Hypothesis 2 : Internal communication positively affects employees' job attitude (which consists of satisfaction and commitment).

3 Methodology

Sampling Procedure

Thirty departments of the University of Jammu were approached for the purpose of data collection using convenience sampling. Questionnaires were distributed to non-teaching staff members belonging to the rural background working in the same 30 departments of the university. Respondents from senior as well as junior levels were contacted for making assessment about the proclivity and communication opportunities provided by their leaders, that is, the HODs. Supportive non-teaching staff included senior officers, senior assistants, senior librarians, senior head assistants and senior personal assistants, junior officers, junior assistants and junior library assistants. Duly filled questionnaires were returned by 52 employees out of 60 non-teaching rural employees and were considered apt for further analysis.

Measures

A five-point Likert scale was used to measure the study constructs. This scale ranges from 1 = strongly disagree to 5 = strongly agree. The present study has taken the existing validated scale of Vakola and Bouradas (2005) consisting of five items each for measuring study variables, that is, leader's proclivity/attitude, internal communication and job commitment. Sample item for measuring HOD's attitude/proclivity is, 'Top management of the company encourages employees to express their disagreements regarding company issues'. The sample item of internal communication is, 'There is adequate communication between employees and top management of this company'. Furthermore, sample item for measuring the level of employees' job commitment is, 'I am very satisfied with my choice to come and work for this company in comparison with other opportunities I had when I was looking for a job'. One item, 'I would not mind working for a different company if the nature of the job was similar', was amended and reverse worded according to the need for the study. Similarly, the scale of job satisfaction is adopted from the four-item scale of Vakola and Bouradas (2005). The sample items are given in a clubbed form by Vakola and Bouradas (2005): 'To which extent are you satisfied or dissatisfied with the following issues which concern your job in this company? (a) Training (b) Rewards in relation with performance (c) Opportunities for promotion (d) Overall my job in this company'. Moreover, all the items used in the present study are moulded as per requirements of the higher education sector concerned.

Data Analysis

Data is analysed using Partial Least Squares (PLS) research technique, which is a structural equation modelling approach used for measuring and testing factor and composite models. PLS depicts relationship between dependent and independent variables and their manifest variables (Dijkstra and Henseler 2015). Sample size of non-teaching rural employees in the present study is 52, which is sufficient to measure data using PLS technique (Wong 2013).

Three models are prepared for accomplishing study objectives. First model represents overall relationship between the selected variable (Fig. 9.2; overall model). Further, to validate the findings, results are also checked individually for leaders' proclivity and internal communication impact on satisfaction and commitment by preparing two other models. Model A (Fig. 9.2) represents relationship between HOD's proclivity and non-teaching staff's job attitude, while Model B (Fig. 9.2) represents the relationship between internal communication and non-teaching supportive staff's job attitude. Reliability of the constructs is measured through Cronbach's alpha as recommended by Nunnally and Bernstein (as cited in Henseler et al. 2016, p. 10). In the overall model (Fig. 9.2), the value of Cronbach's alpha for the construct leader's attitude, internal communication, employees' job commitment and job satisfaction is 0.89, 0.89, 0.86 and 0.95, respectively. In the study, the value of Cronbach's alpha for all the study variables is also found to be above 0.80. Furthermore, all the hypothesised relationships (between leader's proclivity, communication opportunities, job satisfaction and job commitment) are tested and evaluated using PLS (Fig. 9.2; overall Model, Model A, Model B).

Hypothesis Testing and Results

Results confirm that leader's proclivity positively predicts employees' job commitment (SRW = 0.49), and employees' job satisfaction (SRW = 0.28). Hence, first hypothesis is accepted (Fig. 9.2; Model A). The second hypothesis that the impact of internal communication positively affects employees' job commitment (SRW = 0.42) and employees' job

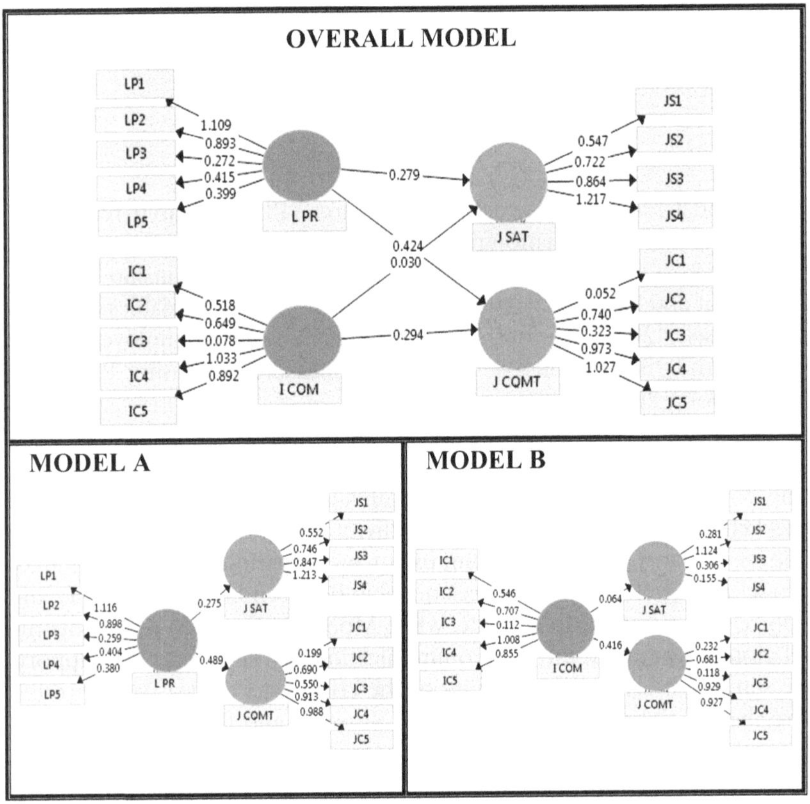

Fig. 9.2 PLS results. (Notes: L PR—Leader's Proclivity, I COM—Internal Communication, J SAT—Job Satisfaction, J COMT—Job Commitment. LP1, LP2, LP3, LP4, LP5 are the indicators of Leader's Proclivity; IC1, IC2, IC3, IC4, IC5 are the indicators of Internal Communication; JS1, JS2, JS3, JS4 are the indicators of Job Satisfaction; JC1, JC2, JC3, JC4, JC5 are the indicators of Job Commitment. L PR and I COM are the predictors of J SAT and J COMT. Source: Author's own findings)

satisfaction (SRW = 0.064) is also accepted (Fig. 9.2; Model B). Moreover, the results of integrated overall model (Fig. 9.2) support that leader's positive proclivity/attitude and internal communication positively predicts employees' job attitude (commitment and satisfaction). Results are consistent for integrated overall model, as well as model (a) and (b). Thus, overall hypothesised relationships are confirmed (Fig. 9.2).

4 Discussion

The first objective of the study is to evaluate the impact of higher authorities' proclivity on rural non-teaching staff's job commitment and satisfaction. The second objective aims to examine the impact of internal communication on non-teaching supportive members' commitment and job satisfaction; the same is reflected in the study models (overall, A and B).

Using PLS technique, beta values of all the study constructs are evaluated. The first hypothesis, which suggested that the leader's proclivity positively affects employees' job attitude (satisfaction and commitment) is accepted and is also supported by previous studies such as the ones done by Thomas et al. (2006), Van Vuuren et al. (2007), Ampofo (2016) and Kim and Beehr (2018). The positive results reflect the presence of commitment and satisfaction from the present job among the rural non-teaching staff members in the higher education sector. All employees contacted in this survey feel proud to be associated and to work in their respective departments of University of Jammu. Moreover, they are also not interested in changing their jobs even if they get an opportunity to work in departments of other organisations. These results imply that rural employees are content with the proclivity of their leaders (higher authorities) in terms of encouragement, treatment responsiveness and so on. The reason for positive commitment and satisfaction might be that rural employees' expectations and requirements are being given due consideration by the higher authorities.

The result of the second hypothesis also highlights the positive impact of internal communication on employees' job attitude, that is, satisfaction and commitment (Fig. 9.2). The findings are in line with previous studies (Thomas et al. 2006; Madlock 2008; Carriere and Bourque 2009; Men 2014; Jacobs et al. 2016). But more specifically, results reveal that rural employees feel more committed towards the organisation than satisfied with regard to communication opportunities received in their present job. The reason for that could be that employees' informal needs and aspirations regarding interactions, sharing, informal encouragement, and so on, are taken care of by their leader, that is, higher authority, which is good enough for rural employees to remain more committed to the institution than be internally satisfied. Another possible reason could be that public educational institutions are more formalised in their nature and job activities of non-teaching staff members who are likely to be concerned with following the same routine every day. This provides less scope and opportunity for formal discussions with higher authorities. Moreover, meetings are frequently arranged for discussions with teaching staff rather than non-teaching ones. Therefore, it can be concluded that in spite of low satisfaction with regard to communication opportunities among non-teaching staff members, higher authorities are able to procure good level of commitment from their members as leaders' attitude remains positive towards them.

5 Managerial Implications

The chapter begins with analysing the role of leaders in influencing employees' workplace attitude. The different types of responses and proclivity/attitude of leaders towards employees' inputs (views, suggestions and problems, etc.) as suggested by various researchers are also discussed and the impact of the leader's (HOD's) proclivity and internal communication on employees' (non-teaching) job satisfaction and commitment are analysed. By empirically examining the research objectives in question, the present study concludes that higher authorities' (HOD's) responsiveness depicting support and regard for non-teaching supportive staff's valuable feedback and important information positively affect their level of satisfaction and commitment. Further, internal communication initiatives in the form of voicing platforms and opportunities by the higher authorities also improve their overall job attitude (satisfaction and commitment). Therefore, the present study also recognises the importance of leader's role in changing followers' job attitude. In organisations where leaders adopt a negative attitude by behaving in a deaf and dumb manner when employees want to have a word with them becomes a negative reinforcement. Further, lack of communication opportunity magnifies their negative perception about the organisational environment and teaches them to apply self-control and monitors, which ultimately affect their job satisfaction and commitment. Higher authorities should create special provisions that ensure maximum participation of employees in the form of their free expression of voice.

HODs should not promote negative beliefs and thoughts in the institute, for instance, the bias that employees at lower level are troublemakers and they cannot be relied upon. Moreover, higher authorities should not force their decisions upon their staff members by discarding important views and suggestions received from them. In addition, those at higher positions should not unnecessarily be suspicious regarding the loyalty of organisational members working under them. At times when formal discussions are arranged for teaching as well as non-teaching staff members, higher authorities should not threaten those who offer their disagreement over the departmental issues. To improve internal communication system in the institute, formal and informal discussions should be encouraged to assure employees that their participation in the form of speaking up will not result in any negative consequences for them. HODs must give adequate information to teaching as well as non-teaching staff on a regular basis. Changes in the department should be brought to the notice of non-teaching staff members together with teaching staff. Further, to improve employees' satisfaction and commitment towards their job and the organisation, higher authorities should make sure that their subordinates are able to relate well with their values and the department's values. Moreover, efforts should be made by the HODs to provide a conducive work environment that provides adequate training opportunities, rewards, and so on, for improving the overall satisfaction of the subordinates. Such initiatives by the higher authorities will fade away the chances that induce employees to quit present institution and repent over their decision to work.

6 Limitations and Future Research

The present study was conducted in the higher education sector among non-teaching rural staff members in India; its results cannot be generalised to other sectors. In future, more multinational studies can be conducted. Future research can also investigate leader's proclivity towards subordinates at all the levels of organisational hierarchy in different sectors. Another limitation of the study is its cross-sectional nature. Longitudinal research design can be followed in future. In this research, we have focused on a few organisational factors, that is, higher authority's (HOD's) proclivity and internal communication that affects non-teaching support staff's job attitude (job satisfaction and commitment). Moreover, the present study is limited to non-teaching staff members. Therefore, it provides opportunity to future researchers to explore more factors affecting employees' job proclivity. In addition, future research can also identify and examine more components of employees' job attitude. Future research can also incorporate more consequences, moderators and mediators between leader's proclivity, communication opportunities and employees' job attitude such as ambidextrous leadership where leader acts according to the demand of the situation.

Appendix

Leader's Proclivity (L PR)

LP1. Your HOD encourages employees (non-teaching staff) to express their disagreements regarding department issues.

LP2. In this department, employees (non-teaching staff) feel free to express themselves.

LP3. If you disagree on departmental issues, you will not be considered as troublemaker by the HOD.

LP4. If you express your disagreement regarding departmental issues you may not suffer negative results from the HOD.

LP5. If you disagree on departmental issues, it will not be considered as lack of loyalty by the HOD.

Internal Communication (I COM)

IC1. Communication with colleagues from other departments is satisfactory in terms of getting a response.

IC2. There is a systematic and organised exchange of knowledge and experiences among non-teaching staff members in this department.

IC3. There is adequate communication between non-teaching staff members and HOD of this department.

IC4. Changes in the department are communicated adequately to the non-teaching staff members.

IC5. The main members in the department keep non-teaching staff informed regarding its mission, plans and progress. You prepare your employees (non-teaching staff) to perform well.

Job Commitment (J COM)

JC 1. You would not like to shift, if you get an opportunity to work in other departments.

JC2. You believe that the department's values and your values are similar.

JC 3. You feel proud to say that you work in the present department.

JC 4. This department encourages you to put maximum effort in order to give more output.

JC 5. You are highly satisfied with your choice to work for this department in comparison with other opportunities you had before looking for this job.

Job Satisfaction (J SAT)

JS 1. You are satisfied regarding training opportunities given in the department.

JS 2. You are satisfied with the rewards/appreciations given in comparison to your performance.

JS 3. You are satisfied with the opportunities given for promotion.

JS 4. You are satisfied with your overall job in the department.

References

Allard-Poesi, F., & Hollet-Haudebert, S. (2017). The Sound of Silence: Measuring Suffering at Work. *Human Relations, 70*(12), 1442–1463.

Ampofo, E. Y. (2016). Immediate Supervisors' Leadership Behaviour and Employees' Organizational Commitment: Do Pay and Promotion Mediate the Nexus? *International Journal of Business and Social Research, 6*(8), 61–72.

An, F., & Bramble, T. (2017). Silence as a Survival Strategy: Will the Silent Be Worse Off? A Study of Chinese Migrant Workers in Guangdong. *The International Journal of Human Resource Management*, 1–26. https://doi.org/10.1080/09585192.2017.1328609.

Asrar-ul-Haq, M., & Kuchinke, K. P. (2016). Impact of Leadership Styles on Employees' Attitude Towards Their Leader and Performance: Empirical Evidence from Pakistani Banks. *Future Business Journal, 2*(1), 54–64.

Babin, B. J., & Boles, J. S. (1996). The Effects of Perceived Co-worker Involvement and Supervisor Support on Service Provider Role Stress, Performance and Job Satisfaction. *Journal of Retailing, 72*(1), 57–75.

Bambacas, M., & Patrickson, M. (2008). Interpersonal Communication Skills that Enhance Organizational Commitment. *Journal of Communication Management, 12*(1), 51–72.

Carriere, J., & Bourque, C. (2009). The Effects of Organizational Communication on Job Satisfaction and Organizational Commitment in a Land Ambulance Service and the Mediating Role of Communication Satisfaction. *Career Development International, 14*(1), 29–49.

Chen, H. C., Beck, S. L., & Amos, L. K. (2005). Leadership Styles and Nursing Faculty Job Satisfaction in Taiwan. *Journal of Nursing Scholarship, 37*(4), 374–380.

Dedahanov, A. T., Lee, D., Rhee, J., & Yusupov, S. (2016). An Examination of the Associations Among Cultural Dimensions, Relational Silence and Stress. *Personnel Review, 45*(3), 593–604.

Dijkstra, T. K., & Henseler, J. (2015). Consistent Partial Least Squares Path Modelling. *Management Information Systems Quarterly, 39*(2), 297–316.

Erkutlu, H., & Chafra, J. (2019). Leader Machiavellianism and Follower Silence: The Mediating Role of Relational Identification and the Moderating Role of Psychological Distance. *European Journal of Management and Business Economics, 28*(3), 20.

Guenter, H., Schreurs, B., Van Emmerik, I. J., & Sun, S. (2017). What Does It Take to Break the Silence in Teams: Authentic Leadership and/or Proactive Followership? *Applied Psychology, 66*(1), 49–77.

Hatipoglu, B., & Inelmen, K. (2017). Demographic Diversity in the Workplace and Its Impact on Employee Voice: The Role of Trust in the Employer. *The International Journal of Human Resource Management*, 1–25. https://doi.org/10.1080/09585192.2017.1322120.

Henseler, J., Hubona, G., & Ray, P. A. (2016). Using PLS Path Modeling in New Technology Research: Updated Guidelines. *Industrial Management and Data Systems, 116*(1), 2–20.

Huang, X., Vliert, E. V., & Vegt, G. V. (2005). Breaking the Silence Culture: Stimulation of Participation and Employee Opinion Withholding Cross-Nationally. *Management and Organization Review, 1*(3), 459–482.

Jacobs, M. A., Yu, W., & Chavez, R. (2016). The Effect of Internal Communication and Employee Satisfaction on Supply Chain Integration. *International Journal of Production Economics, 171*, 60–70.

Jones, A., & Kelly, D. (2014). Deafening Silence? Time to Reconsider Whether Organization Are Silent or Deaf When Things Go Wrong. *BMJ Quality and Safety, 23*(9), 709–713.

Kim, M., & Beehr, T. A. (2018). Empowering Leadership: Leading People to Be Present Through Affective Organizational Commitment? *The International Journal of Human Resource Management*, 1–25. https://doi.org/10.1080/09585192.2018.1424017.

Kish-Gephart, J. J., Detert, J. R., Trevino, L. K., & Edmondson, A. C. (2009). Silenced by Fear: The Nature, Sources, and Consequences of Fear at Work. *Research in Organizational Behavior, 29*, 163–193. https://doi.org/10.1016/j.riob.2009.07.002.

Madlock, P. E. (2008). The Link Between Leadership Style, Communicator Competence, and Employee Satisfaction. *The Journal of Business Communication, 45*(1), 61–78.

Men, L. R. (2014). Strategic Internal Communication: Transformational Leadership, Communication Channels, and Employee Satisfaction. *Management Communication Quarterly, 28*(2), 264–284.

Morrison, E. W., & Milliken, F. J. (2000). Organizational Silence: A Barrier to Change and Development in a Pluralistic World. *Review Academy of Management, 25*(4), 706–725.

Nobile, J. (2017). Organisational Communication and Its Relationships with Job Satisfaction and Organisational Commitment of Primary School Staff in Western Australia. *Educational Psychology, 37*(3), 380–398.

Orpen, C. (1997). The Interactive Effects of Communication Quality and Job Involvement on Managerial Job Satisfaction and Work Motivation. *The Journal of Psychology, 131*(5), 519–522.

Park, J. H., Carter, M. Z., DeFrank, R. S., & Deng, Q. (2016). Abusive Supervision, Psychological Distress, and Silence: The Effects of Gender Dissimilarity Between Supervisors and Subordinates. *Journal of Business Ethics*, 1–18. https://doi.org/10.1007/s10551-016-3384-3.

Pinder, C. C., & Harlos, K. P. (2001). Employee Silence: Quiescence and Acquiescence as Responses to Perceived Injustice. In K. M. Rowland & G. R. Ferris (Eds.), *Research in Personnel and Human Resources Management* (Vol. 20, pp. 331–369). New York: JAI Press.

Ruppel, C. P., & Harrington, S. J. (2000). The Relationship of Communication, Ethical Work Climate, and Trust to Commitment and Innovation. *Journal of Business Ethics, 25*(4), 313–328.

Srivastava, P., & Frankwick, G. L. (2011). Environment, Management Attitude and Organizational Learning in Alliances. *Management Decision, 49*(1), 156–166.

Thomas, W., Butts, M. M., Vandenberg, R. J., Joy, D. M., & Wilson, M. G. (2006). Effects of Management Communication, Opportunity for Learning, and Work Schedule Flexibility on Organizational Commitment. *Journal of Vocational Behavior, 68*(3), 474–489.

Vakola, M., & Bouradas, D. (2005). Antecedents and Consequences of Organisational Silence: An Empirical Investigation. *Employee Relations, 27*(5), 441–458.

Van Dyne, L., Ang, S., & Botero, I. C. (2003). Conceptualizing Employee Silence and Employee Voice as Multidimensional Constructs. *Journal of Management Studies, 40*(6), 1359–1392.

Van Vuuren, M., Jong, M. D., & Seydel, E. R. (2007). Direct and Indirect Effects of Supervisor Communication on Organizational Commitment. *Corporate Communications: An International Journal, 12*(2), 116–128.

Venkataramani, V., Zhou, L., Wang, M., Liao, H., & Shi, J. (2016). Social Networks and Employee Voice: The Influence of Team Members' and Team Leaders' Social Network Positions on Employee Voice. *Organizational Behavior and Human Decision Processes, 132*, 37–48. https://doi.org/10.1016/j.Obhdp.2015.12.001.

Willman, P., Bryson, A., & Gomez, R. (2006). The Sound of Silence: Which Employers Choose 'no voice' and Why? *Socio-Economic Review, 4*(2), 283–299.

Wong, K. K. (2013). Partial Least Squares Structural Equation Modeling (PLS-SEM) Techniques Using Smart PLS. *Marketing Bulletin, 24*, 1–32.

Xu, A. J., Loi, R., & Lam, L. W. (2015). The Bad Boss Takes It All: How Abusive Supervision and Leader–Member Exchange Interact to Influence Employee Silence. *The Leadership Quarterly, 26*(5), 763–774.

Yousef, D. A. (2000). Organizational Commitment: A Mediator of the Relationships of Leadership Behavior with Job Satisfaction and Performance in a Non-western Country. *Journal of Managerial Psychology, 15*(1), 6–24.

CHAPTER 10

Impact of Self-efficacy on Entrepreneurial Intentions: Role of Self-regulation and Education

Sumeet Kour and Mamta Sharma

1 Introduction

Entrepreneurship presents an important component of dynamism into an economic system. Entrepreneurship in India is led by the communal system of the ancient past, that is, Brahmins helped Kshatriyas in the management, vaishyas have achieved industrial productive activities and trading and shudras are involved in agrarian business (Swath and Rao 2013). The people are organised in a very modest kind of social and economic system. In the way to implement this concept to modern entrepreneurship, it can compare with the villages as organisation and an entrepreneur as a craftsman (Swath and Rao 2013). Entrepreneurship is core development and is fundamentally a creative activity. It is considered as one of the more important factors for industrial growth and has significant contribution in economic development of the society (Tseng 2012). Entrepreneurship and strategic management

S. Kour
Cluster University of Jammu, Jammu, Jammu and Kashmir, India

M. Sharma (✉)
Udhampur Campus, University of Jammu, Jammu, Jammu and Kashmir, India

H. Chahal et al. (eds.), *Sustainable Business Practices for Rural Development*, https://doi.org/10.1007/978-981-13-9298-6_10

are concerned with growth and wealth creation (Amit and Zott 2001). Growth and wealth creation are the defining objectives of entrepreneurship (Kuratko et al. 2005). Entrepreneurs are the backbone of the economy. They are the moving force behind any economy and are invaluable in bringing about a dynamic change in the economy. It is also an accepted belief that without entrepreneurial activities the process of development is not possible (Audretsch et al. 2002). Real entrepreneurs are zealous, practical and motivated to prosper. An entrepreneur not only perceives and organises business but also takes risks while doing so. A nation's economic development is based on the industrial development and it is constructed on the entrepreneurial skills of the people. Therefore, entrepreneurship advancement is the need of the hour (Santhi and Kumar 2011). Entrepreneurship accelerates the progress of regions, nations and communities. Entrepreneurs keenly endorse entrepreneurship development, reveal greater growth rates and subsequently higher levels of development. Religious belief and culture significantly influence the choice of an individual to engage in entrepreneurial activities and economic behaviour in society (Meyer 2009). Small-scale entrepreneurship plays an important role in the economic development of a developing country like India. Entrepreneurship is not only vital to solve the problems of industrial growth but also to solve problems related to unbalanced regional development, unemployment and concentration of economic power (Latha et al. 2008). The small savings of rural areas are backing the micro and small industries in India and other developing countries such as Sri Lanka, Nepal, Bhutan and Myanmar. From this perspective, rural entrepreneurship has been given much importance (Vinesh 2014).

2 Rural Entrepreneurship in India

Entrepreneurship has garnered much importance at the international level under varying economic settings. International economy, and specifically the Indian economy, is growing at a faster rate as far as entrepreneurship is concerned. Super mall culture provides the scope for entrepreneurship in trading and manufacturing. Rural entrepreneurship refers to entrepreneurship emerging in rural areas (Jayadatta 2017). Hence, rural entrepreneurship is the same as rural industrialisation, as both focus on establishing small-scale industries in rural areas and developing people living in such areas. Rural entrepreneurship has a vital role in the development of the Indian economy. About 70 per cent of the Indian population calls rural India their home and proper funding and support gives entrepreneurs an adequate environment in these rural societies to nurture their businesses

(Selladurai and Priyadharshini 2015). Rural entrepreneurship is a key prospect for individuals who immigrate from rural and semi- urban areas to urban areas in search of employment. Majority of rural entrepreneurs face several difficulties because of non-availability of basic amenities in rural areas, particularly in India (Jayadatta 2017). Financial problems, lack of education, inadequate technical and theoretical ability make it difficult for rural entrepreneurs to create businesses in rural areas.

One of the key objectives of growth plans in India is to give employment to many unemployed youths and especially those that reside in rural areas. The fundamental difficulty for countries like India is the excess of agricultural labour and shutting down of traditional village industries, resulting in greater unemployment in rural areas and relocation of rural youth to urban areas in desperate hunt for jobs which has further put pressure on the urban amenities and infrastructure. Rural enterprises created employment for over 47.97 lakhs individuals in the year 1996–1997, as compared to 37.21 lakhs individuals in the year 1992–1993. There are a large number of products and services present in rural areas which can be leveraged by entrepreneurs to start new macro and micro enterprises. There is significant growth in rural development after the 14th Finance Commission recommendations. However, the outcome was not very satisfactory. Only 20 per cent or less benefits from the government schemes have reached the poor and hence the total development of rural people itself is low (Vedanthadesikan and Pathmanathan 2016). Therefore, there is a need to look into why poverty, under-development and unemployment continue to exist in most parts of the country, and particularly in rural areas.

Regardless of the insufficiencies in the rural areas, we must consider the strengths and make rural areas a place of opportunities (Patel and Chavda 2013). The rural youth analyse things differently as they have seen more challenging situation to achieve their objectives (Patel and Chavda 2013). It doesn't mean that there is a shortage of individuals with such mindset. But with changing times they change their minds and join the trend of job seekers as they are unaware about the opportunities prevailing in the market due to lack of information. Youth with such type of mindset and with the help of correctly channelised drives can succeed in the era of rural entrepreneurship (Patel and Chavda 2013).

In the rural areas, entrepreneurship is a good source for providing better job opportunities for the youth and is a sustainable way to integrate both international and domestic labour markets (ILO 2009; Salkowitz 2010). Therefore, the purpose of the study is to examine the entrepreneurial intentions of the rural youth (Fig. 10.1).

3 Theoretical Framework

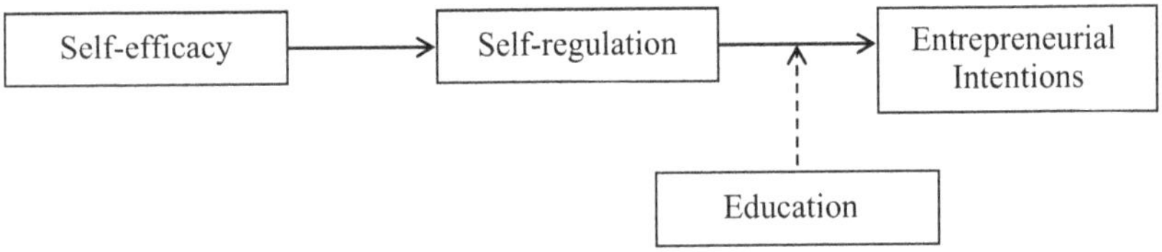

Fig. 10.1 Conceptual model. (Source: Authors)

4 Literature Review and Hypothesis Development

Self-efficacy and Self-regulation

Self-efficacy is the conviction that one can successfully execute the desired behaviour required to produce an outcome (Bandura 1986). It is the individual's judgements about their own capability to accomplish a specified activity (Campo 2011). Self-efficacy impacts the strength one puts forth to alter risk behaviour and the persistence to continue striving in spite of barriers and setbacks that may weaken motivation. This phenomenon positively affects self-regulation (Ghonsooly and Ghanizadeh 2013; Pihie and Bagheri 2013; Schunk and Zimmerman 2007). Bandura (1997) revealed that self-efficacy affects one's behaviour through self-regulatory processes such as motivation to pursue and follow personal standards and goals, willpower and self-confidence in successfully performing a job. On the same lines, Tumasjan and Braun (2012) revealed that there is a significant association between self-efficacy, self-regulation and entrepreneurs' ability to recognise innovative and creative opportunity to start a venture. Self-efficacious individuals stick to their aims and accomplish their task, which in turn helps them regulate themselves. They find different ways to solve problems and learn from their mistakes. Further, Pihie and Bagheri (2013) have revealed that students who are self-efficacious have high self-regulation as they can deal efficiently with unexpected events and handle unpredictable situations successfully. Therefore, it can be determined from the earlier discussion that self-efficacy positively affects self-regulation.

Hypothesis 1 : Self-efficacy positively affects self-regulation.

Self-Regulation and Entrepreneurial Intention

Self-regulation is an organised method of an individual's thought and behaviour that includes setting individual goals and directing one's self towards the accomplishment of those goals (Bryant 2006). It is a process that consists of observing, assessing and providing response on individual actions through self-reinforcement and self-control to transmit actions towards attaining objectives (Bandura 2006; Bendassolli et al. 2016). The purpose of self-regulation is to help a person to take individual control of the settings in which he/she is positioned, enhancing his/her ability as an agent (Bendassolli et al. 2016; Forgas et al. 2009; Mezo 2008). Studies have revealed that self-regulation positively affects entrepreneurial intentions (Hu et al. 2017; Pihie and Bagheri 2013). Self-regulation plays a vital role in one's selection as an entrepreneur (Bryant 2006) and it also affects the amount of effort entrepreneurs put into starting new ventures as well as the entrepreneurial success (Brockner et al. 2004). Self-regulated individuals can take personal control of the environment which encourages them to set up their own venture. Brockner et al. (2004) viewed that the complex and challenging decision to become an entrepreneur and the subsequent intention to explore new business ideas require a high sense of regulatory focus. Further, they revealed that with a moderate level of self-regulation, individuals are less likely to explore novel business ideas and step into the challenging course of creating new ventures. Self-regulation stimulates the intentions of the individuals to start their own venture. Self-regulated individuals have rich understanding of entrepreneurship (Baron et al. 2016; Frese 2009). Self-regulation motivates entrepreneurs to explore numerous creative and innovative entrepreneurial opportunities (Brockner et al. 2004; Trevelyan 2011) and also have to choose which entrepreneurial opportunities to exploit (Bryant 2007) and improve the performance of their new enterprise (Hmieleski and Baron 2008). Bryant Peter (2009) revealed that entrepreneurs with more self-regulatory features are ethically aware and relate such consciousness to maintain individual integrity and build strong level of interpersonal trust which motivates them to start their own venture. Hence, it can be concluded that self-regulated individuals have high entrepreneurial intentions.

Hypothesis 2 : Self-regulation positively affects entrepreneurial intentions.

Self-efficacy, Self-regulation and Entrepreneurial Intentions

Self-regulation is controlling, managing and organising oneself without any external intervention in order to reach the preferred goal. In other words, if one regulates himself/herself then they can manage, organise and control the business successfully. The person with high self-regulation becomes a successful entrepreneur. This phenomenon mediates between self-efficacy and entrepreneurial intentions (Pihie and Bagheri 2013). Individuals who are confident about their abilities and skills are able to control their behaviour more, which in turn motivates them to engage themselves in entrepreneurial activities or set up their own ventures. In order to start and run a business lucratively, there is need for individual's adaptability to the situation, which can be possible only with the help of regulation. Self-efficacious individuals are able to manage difficult situations and learn from their mistakes, which in turn helps them to undertake entrepreneurship. They set their personal standards and follow them strictly, which in turn enhances their entrepreneurial intentions. Self-efficacious individuals have the capability to successfully complete the specific task (Paulsen and Gentry 1995; Schunk 1996; Onoda 2014), which in turn helps individual to regulate themselves thereby improving their entrepreneurial intentions. Self-efficacy fuels motivation (i.e., a drive to initiate their learning) and volition (i.e., willpower to help individuals to maintain their learning when they are faced with distracting factors), enabling individuals to persist when faced with difficulties (Onoda 2014) which further leads to higher entrepreneurial intentions. Therefore, it can be concluded from this discussion that self-regulation mediates the relationship between self-efficacy and entrepreneurial intentions.

Hypothesis 3 : Self-regulation mediates between self-efficacy and entrepreneurial intentions.

Self-regulation, Education and Entrepreneurial Intentions

Self-regulation is a methodical procedure of individual thought and behaviour that includes setting individual goals and directing oneself towards the accomplishment of those goals (Bryant 2006). Studies have revealed that self-regulation positively affects entrepreneurial intentions (Hu et al. 2017; Pihie and Bagheri 2013). Self-regulation plays an essential part in selection of an individual as an entrepreneur (Bryant 2006) and it also affects the amount of effort entrepreneurs put into starting new ventures as well as their entrepreneurial success (Brockner et al. 2004). The association between self-regulation and entrepreneurial intentions strengthens when the individuals have the education or knowledge about entrepreneurship. Entrepreneurship education is seen as a means of nurturing the entrepreneurial spirit and behaviour of people (Venesaar et al. 2011). Entrepreneurship education creates real-life time pressured learning environments with unexpected events (Cope 2003; Gibb 2008; Pittaway and Cope 2007) which strengthen the effect of self-regulation on entrepreneurial intentions. Individuals, who are self-regulated and have entrepreneurial education, have high motivation to set up their own venture. Self-regulated individuals with entrepreneurial education facilitate the occurrence of new venture and the growth of person's business skills, which makes it likely to increase achievement in both work and personal life (Ling and Venesaar 2015). Individuals with high self-regulation and entrepreneurial education have more entrepreneurial intentions as they have the capability as well as the required knowledge to set up their own business. Individuals who are aware of entrepreneurship concepts, its importance and the various opportunities available to them are able to plan, manage and control their entrepreneurial activities thereby enhancing entrepreneurial intentions. Therefore, it can be concluded that the relationship between self-regulation and entrepreneurial intentions get strengthened when individuals have entrepreneurial education.

Hypothesis 4 : Education moderates the relationship between self-regulation and entrepreneurial intentions.

5 Method

To make the present study more objective, the following steps have been taken.

Data Collection

The data for the study has been gathered from undergraduate, graduate and postgraduate students studying in various colleges and universities in Udhampur district (Jammu and Kashmir). In all, 350 questionnaires were distributed, of which only 251 responded. Purposive sampling has been used to collect the data. Only those students were contacted who belong to the rural areas of Jammu and Kashmir. To collect the data, a questionnaire has been used. To establish the normality of the data, 26 respondents have been deleted by inspecting boxplots. The retained data exhibited normalcy, as skewness (−0.714) and Kurtosis (1.054) are within the range. Therefore, the effective sample came to 225.

Measures

To measure the variables used in the present study, a five-point Likert scale has been used, ranging from strongly disagree (1) to strongly agree (5).

Self-efficacy: It has been measured with the help of eight items (Schwarzer and Jerusalem 1995).
Self-regulation: 15 items have been used to measure self-regulation (Brown et al. 1999).
Entrepreneurial Intentions: Six items have been used to measure entrepreneurial intentions (Lüthje and Franke 2003).
Education: Respondents have been asked if they 'studied entrepreneurship as a subject'.
Control Variables: Age and gender are the control variables used in the study.

6 Results

Exploratory Factor Analysis (EFA)

To find the dimensions of different scales used in the present study EFA has been conducted. Vari-max rotation in principle component analysis has been used. KMO measure has been used to check the sampling adequacy, where values above 0.50 are tolerable (Hair et al. 2010), indicating its relevance for advance analysis. The items with factor going below 0.50 have been deleted (Hair et al. 2010). The entrepreneurial intentions scale consisted of six items that got reduced to four items which converged under one factor. Similarly, self-efficacy scale initially consisted of eight items that got reduced to four items and converged under one factor. Further, the self-regulation scale consisted of 15 items which has been reduced to four items and converged under single factors. The Kaiser-Meyer-Olkin Measure of Sampling Adequacy (KMO) value of all the constructs is above 0.80 and total variance explained for all the constructs is above 80 per cent (Hair et al. 2010) according to Table 10.1.

Table 10.1 Results of exploratory factor analysis

Factor	*M*	*SD*	*FL*	*C*	*E.V*	*V.E. (%)*	*KMO*	*Cronbach's alpha*
Entrepreneurial intentions	*3.66*	*0.77*			*2.412*	*60.290*	*0.770*	*0.780*
EI1	3.78	0.96	0.79	0.63				
EI2	3.61	0.84	0.79	0.63				
EI4	3.42	0.96	0.78	0.61				
EI5	3.84	0.97	0.74	0.55				
Self-efficacy	*3.61*	*0.72*			*2.541*	*63.525*	*0.741*	*0.728*
SE3	3.56	1.04	0.74	0.55				
SE4	3.54	1.08	0.75	0.57				
SE5	3.60	0.91	0.74	0.56				
SE8	3.72	0.82	0.74	0.54				
Self-regulation	*3.82*	*0.61*			*2.456*	*61.423*	*0.763*	*0.761*
SR4	3.72	0.85	0.77	0.60				
SR12	3.95	0.76	0.78	0.61				
SR13	3.80	0.82	0.73	0.53				
SR14	3.83	0.75	0.77	0.59				

Source: Authors

Notes: SD = Standard Deviation, M = Mean, FL = Factor Loading, E.V = Eigen Value, C = Communality, KMO = Kaiser-Meyer-Olkin Measure of Sampling Adequacy, V.E = Variance Explained

Confirmatory Factor Analysis (CFA)

Validity and reliability of all the constructs have been determined by conducting Confirmatory Factor Analysis (CFA). Furthermore, to assess the hypotheses SEM (Structural Equation Modelling) has been used.

For all the scales used in the study, measurement models have been prepared as single factor emerged after EFA. Items with standardised regression weight (SRW), that is, less than 0.50, have been deleted (Hair et al. 2010). Since the SRW for all the items are greater than 0.50 in CFA, no item has been deleted. All the measurements model fit indices are in the threshold limit (Table 10.1). The prescribed values of goodness of fit indices like GFI (Goodness of Fit Index), CFI (Comparative Fit Index) and AGFI (Adjusted Goodness of Fit Index) and the inadequacy of fit criteria like RMSEA (Root Mean Square Error of Approximation) and RMR (Root Mean Square Residual) should be below 0.80 and 0.50, respectively, as recommended by Hair et al. (2010). SRW (greater than 0.60) and AVE (average variance extracted) which is greater than 0.50 proved the convergent validity (Table 10.2). The discriminant validity is

Table 10.2 Validity, reliability and model fit indices

Constructs	*M*	*SD*	*SRW*	*AVE*	*CR*	*Cronbach's alpha*	*Model fitness*
Entrepreneurial intentions	*3.66*	*0.77*		*0.88*	*0.97*	*0.78*	χ^2/df = 2.861
							GFI = 0.987
EI1	3.78	0.96	0.72				RMR = 0.027
EI2	3.61	0.84	0.71				AGFI = 0.936
EI4	3.42	0.96	0.69				CFI = 0.968
EI5	3.84	0.97	0.63				RMSEA = 0.081
Self-efficacy	*3.61*	*0.72*		*0.82*	*0.95*	*0.72*	χ^2/df = 2.837
SE3	3.56	1.04	0.63				GFI = 0.988
SE4	3.54	1.08	0.65				RMR = 0.027
SE5	3.60	0.91	0.64				AGFI = 0.942
SE8	3.72	0.82	0.62				CFI = 0.979
							RMSEA = 0.083
Self-regulation	*3.82*	*0.61*		*0.91*	*0.97*	*0.76*	χ^2/df = 2.084
SR4	3.72	0.85	0.68				GFI = 0.991
SR12	3.95	0.76	0.70				RMR = 0.015
SR13	3.80	0.82	0.61				AGFI = 0.956
SR14	3.83	0.75	0.64				CFI = 0.989
							RMSEA = 0.070

Source: Authors

Notes: M = Mean, SD = Standard Deviation, SRW = Standardised regression weight, AVE = Average variance explained, CR = Composite reliability

proved by relating the AVE values with squared correlations amongst different scales (Fornell and Larcker 1981). The AVE that was extracted from all the scales is more than the squared correlation, thus establishing discriminant validity (Table 10.3). Composite reliability and Cronbach's alpha have been calculated to check the reliability of the constructs. Alpha values equal to or above 0.70 represent good reliability (Hair et al. 2010). The composite reliability and alpha values for all the scales are above 0.70 (Table 10.2), which represents that the scales are reliable.

Common Method Bias (CMB)

The data has been collected from a single respondent only, which can inflate the relationship; therefore, common method bias has been examined through Harman's −1 factor method (Hair et al. 2010). The findings revealed that the total variance explained by one factor is only 23 per cent which is less than 50 per cent, indicating that CMB is not the issue in the study.

Table 10.3 Discriminant validity and correlation analysis

Constructs	*Entrepreneurial intentions*	*Self-efficacy*	*Self-regulation*
Entrepreneurial intentions	*0.88*		
Self-efficacy	(0.030) 0.176**	*0.91*	
Self-regulation	(0.030) 0.176**	(0.076) 0.277**	*0.89*

Source: Authors

Note: Values on the diagonal axis represents the average variance extracted. Values below the diagonal axis are correlation and values in the parentheses represent the squared correlation. $^{**}p < 0.01$

Hypotheses Testing

In order to check the hypotheses, SEM has been applied. In the present study, the relationship between self-efficacy, self-regulation, entrepreneurial intentions and locus of control have been assessed.

We first assessed the effect of self-efficacy on self-regulation and the analysis revealed that self-efficacy significantly affects self-regulation (SRW = 0.37, $p < 0.001$ according to Fig. 10.2). Hence, Hypothesis 3 stands proved.

We assessed the effect of self-regulation on entrepreneurial intentions and the findings indicated that self-regulation positively affects entrepreneurial intentions (SRW = 0.22, $p < 0.01$ according to Fig. 10.3), which confirmed Hypothesis 2. Furthermore, the model yielded a good fit (CMIN/DF = 2.646, RMR = 0.049, GFI = 0.955, AGFI = 0.905, CFI = 0.942, RMSEA = 0.086).

Mediation Effect

Preacher and Hayes' (2004) methodology has been used to check the mediation in the present study. They revealed that mediation analysis is based on the significance of tests of the indirect effect and for this the Sobal test (1982) is the most well-known. Hence, in the present study, the estimate of the indirect effect has been measured with the help of Sobal test and bootstrap approach. The findings show that self-efficacy has a significant effect on self-regulation (SRW = 0.38, $p < 0.001$) and that self-regulation also significantly affects entrepreneurial intentions (SRW = 0.23, $p < 0.01$). Bootstrapping results indicated significant indirect effects of self-efficacy on entrepreneurial intentions through self-regulation (SRW = 0.09, $p < 0.05$ according to Table 10.4) with a 95 per cent confidence interval from 0.002 to 0.222. Lastly, the control variables have also

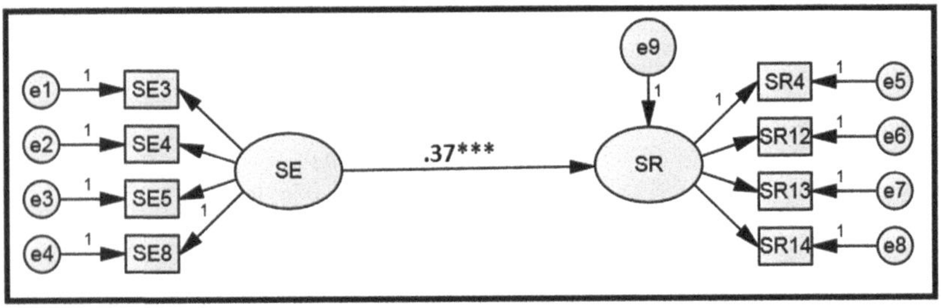

Fig. 10.2 Impact of self-efficacy on self-regulation. (Notes: SR = Self-regulation, SE = Self-efficacy, e1–e9 = error terms. Source: Authors)

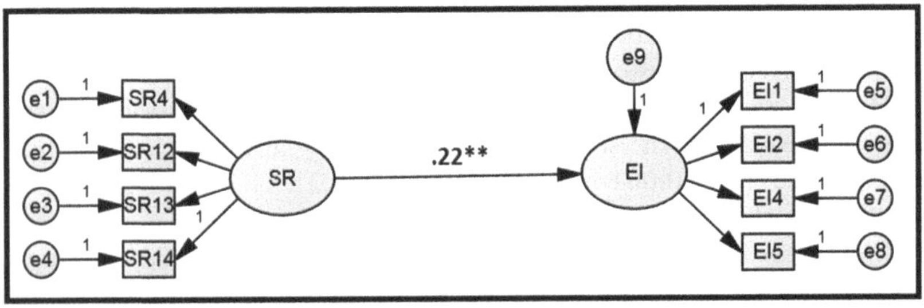

Fig. 10.3 Impact of self-regulation on entrepreneurial intentions. (Notes: SR = Self-regulation, EI = Entrepreneurial intention, e1–e9 = error terms. Source: Authors)

Table 10.4 Bootstrapping results for mediation

Hypothesis	*SE→SR*	*SR→EI*	*Indirect effect*	*LL95% / UL 95%*
SE→SA→EI	0.38***	0.23**	0.09*	CC

Source: Authors

Notes: *** $p < 0.001$, ** $p < 0.01$; $N = 5000$ Bootstrapping resamples; LL BCA and UL BCA = Lower level and Upper level of the bias corrected and accelerated confidence interval. SE = Self-efficacy, SA = Self-regulation, EI = Entrepreneurial intentions

been included in the mediation model, which yielded no change in the earlier relationships, so that is why control variables are not presented in the diagram. Hence, Hypothesis 3 stands proved (Fig. 10.4).

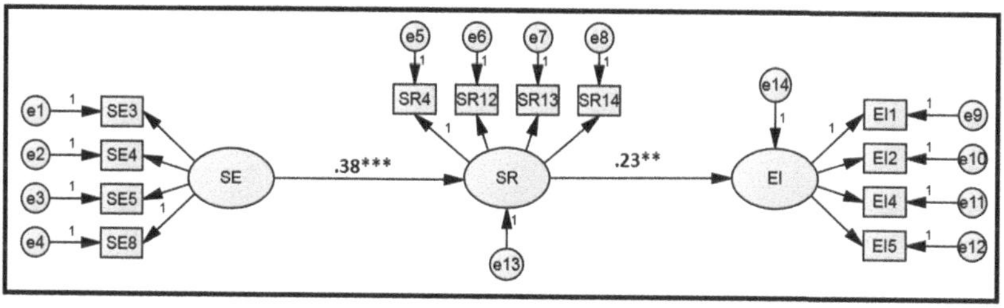

Fig. 10.4 Mediation model. (Notes: SE = Self-efficacy, SR = Self-regulation, EI = Entrepreneurial intention, e1–e14 = error terms. Source: Authors)

Moderating Effect

Structural equation modelling has been used to test the hypotheses in the present study (Byrne 2010). In this study, we have one moderating variable, that is, education. Therefore, moderating effect of education has been checked using a multi-group analysis of SEM. The sample has been divided into two groups. The first group comprises students who have studied entrepreneurship as a subject in their college or university and the second group consists of students who have never studied entrepreneurship as a subject in their college or university, to determine the existence of difference in structural parameters (Jiménez-Jiménez and Sanz-Valle 2011).

In the first step, we have controlled the parameter for the hypothesised relationship to be equal and in the second step, the parameters are not controlled. If the difference between the two models is significant ($\Delta\chi^2$), it indicates that the variable used for dividing the sample moderates the relationship.

Table 10.5 shows that the self-regulation and entrepreneurial intentions' relationship is positive and significant for the first group only, that is, students who have studied entrepreneurship. Thus, education moderates between self-regulation and entrepreneurial intentions' relationship (Table 10.5). Furthermore, the chi-square difference test shows that the two models, that is, constrained and unconstrained models, are different in case of both the groups. The findings are further supported by simple slope analysis (Fig. 10.5). Thus, Hypothesis 4 stands proved.

Table 10.5 Comparison of constrained and unconstrained models for education as moderator

	Education	
	High	*Low*
Self-regulation → Entrepreneurial intentions	0.30*	0.14(ns)
R^2	0.09	0.02
χ^2 constrained model	29.173	86.504
χ^2 unconstrained model	23.869	59.575
$\Delta\chi^2$	5.304***	26.929***

Source: Author

Note: $^{***}p < 0.001$, $^{*}p < 0.05$

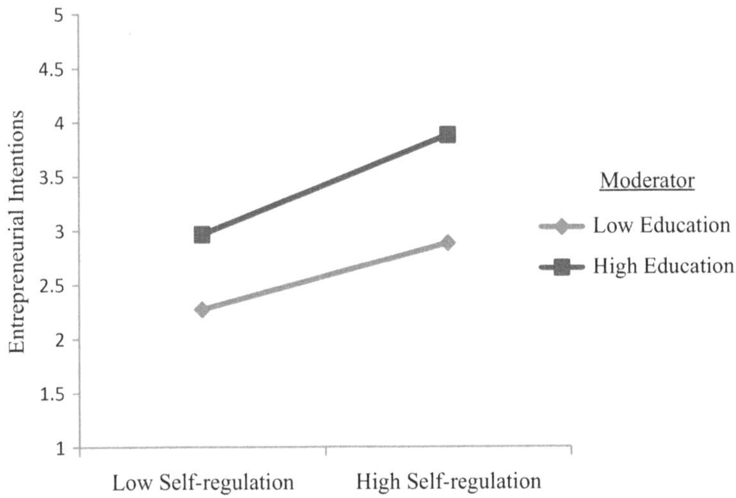

Fig. 10.5 Moderating effect of education. (Source: Author)

Test of Moderated Mediation

The present study has checked the model where the strength of the relationship between self-efficacy on entrepreneurial intentions through self-regulation is dependent on the moderator, which in this case is education. The moderated mediation is confirmed when the indirect effect of self-efficacy on entrepreneurial intentions in the presence of moderating

Table 10.6 Bootstrapped conditional indirect effect of self-efficacy on entrepreneurial intentions through self-regulation with education as value (Moderator)

Moderator	*Level*	*Conditional indirect effect*	*Boot SE*	*Boot LL 95%*	*Boot UL 95%*
Education	High	0.163*	0.108	0.010	0.443
	Low	0.046 (ns)	0.060	−0.022	0.224

Source: Authors

Notes: $^{*}p < 0.05$; $N = 5000$ Bootstrapping resamples; LL BCA and UL BCA = Lower level and Upper level of the bias corrected and accelerated confidence interval

variable is significant. Moderated mediation is proven when the conditional indirect effect of independent variable on dependent variable, via mediator, differs in strength across low and high levels of moderator (Preacher et al. 2007).

The result of Hypothesis 3 revealed that self-regulation mediates between self-efficacy and entrepreneurial intentions' relationship (Table 10.4). Furthermore, the results of Hypothesis 4 revealed that education moderates between self-regulation and entrepreneurial intentions (Table 10.5). The moderated-mediation effect of self-efficacy on entrepreneurial intentions through self-regulation for education (moderators) is significant for high groups, that is, for students who have studied entrepreneurship as a subject in college or university as the bootstrapping results revealed significant indirect effect of self-efficacy on entrepreneurial intentions through self-regulation in presence of high education level. (Table 10.6).

7 Discussion

The present study developed the model that explained the effect of self-efficacy on self-regulation. The study also analysed the impact of self-regulation on entrepreneurial intentions' relationship. The study further examined the mediating role played by self-regulation between self-efficacy and entrepreneurial intentions. Lastly, the study also discusses moderating role played by entrepreneurial education between self-regulation and entrepreneurial intentions.

It can be said that the study revealed that self-efficacy positively affects self-regulation. The results are consistent with the earlier study (Ghonsooly and Ghanizadeh 2013; Pihie and Bagheri 2013; Schunk and Zimmerman 2007). Self-efficacious students stick to their goals and see their assignments, which in turn aids them to regulate themselves. They find diverse ways to resolve problems and learn from their mistakes. Self-efficacious individuals are highly self-regulated as they are creative and accomplish their work in a planned manner. Further, the study revealed that self-regulation positively affects entrepreneurial intentions. The finding is in line with the earlier studies (Hu et al. 2017; Pihie and Bagheri 2013). Self-regulated individuals have personal control of the environment, which motivates them to set up their own business. Self-regulated individuals explore new business ideas and enter into the challenging process of establishing new businesses. Lastly, study also revealed that education moderates the relationship between self-regulation and entrepreneurial intentions. Entrepreneurship education generates real-life time pressured learning environments with unexpected events, which strengthens the impact of self-regulation on entrepreneurial intentions. Individuals who are self-regulated and have entrepreneurial education have more entrepreneurial intentions as they have the capability as well as the required knowledge to set up their own business.

Lastly, the findings of the present study established that self-regulation mediates the combined effect of self-efficacy and entrepreneurial education on entrepreneurial intentions. Moderated mediation is proven when the conditional indirect effect of self-efficacy on entrepreneurial intentions through self-regulation varies in strength across the absence and presence of a moderator, that is, education. The findings revealed that self-regulation mediates the relationship between the interaction of self-efficacy and entrepreneurial education on job performance.

8 Implications

Theoretical Implications

The study confirmed the reliability and constructs validity of the self-efficacy, self-regulation and entrepreneurial intention scales in the Indian context, which increased the generalisability of these scales. It further adds to the self-efficacy–entrepreneurial intentions literature by examining the role played by self-regulation in this relationship. The present study demonstrates that entrepreneurial education strengthens the impact of self-regulation on entrepreneurial intentions. Further, the study examines the integrative model of the relationship between self-efficacy and entrepreneurial intentions with mediating with mediator, that is, self-regulation and moderator, that is, entrepreneurial education. The study has also examined the moderated-mediation of variables (self-regulation and entrepreneurial education) in self-efficacy and entrepreneurial intentions relationship that enhances the entrepreneurship literature.

Practical Implications

The study has several practical implications. It revealed that entrepreneurial education plays an important role in strengthening the relationship between self-regulation and entrepreneurial intentions. Hence, there is a need to regulate entrepreneurial education and training practices. Entrepreneurial education should focus on the development of entrepreneurial skills, capabilities and behaviours. In this sense, development of entrepreneurial potential is possible through a quality system of education. Entrepreneurial course should be designed in such a way that the self-belief of the individuals can be developed. In order to achieve an effective increase in entrepreneurial intentions, it is recommended that there should be guidance into strategies that are more focused on the change of personal attitudes rather than on the theoretical knowledge being taught. Furthermore, to enhance the self-regulation among students, they should be encouraged to create clear goals for themselves and to focus their attention on positive behaviours that must be performed to meet these goals. Students' goals and aspirations should not be so easy to achieve that they don't feel challenged at all. Goals that are too tough will have a similar impact, so job objectives that are moderately challenging will give the best results.

Self-regulation is about remaining calm in the face of adversity and keeping oneself cool-headed. Relaxation techniques such as deep breathing can help students to calm down as it interrupts negative thoughts and puts the focus back on a more positive path. Courses for the students should focus on areas that allow the students to take their own decisions such as assignment options or self-determined due dates, flexible grading and so on. While raising and giving encouragement is very important, it must be reliable too. While giving performance feedback one shouldn't compare one student with another, rather do a comparison with their earlier performance.

9 Limitations and Future Research

All the steps have been taken to ensure the reliability, objectivity and validity of the present study but still there are certain limitations. Firstly, the present study is cross-sectional in nature. So, it is recommended to conduct longitudinal study for more reliable results. Secondly, data has been collected from a single source; in future, data must be gathered from multiple sources for more valid results. Further, more factors affecting entrepreneurial intentions can be included for better understanding of the concept.

References

Amit, R., & Zott, C. (2001). Value Creation in E-business. *Strategic Management Journal, 22*(2), 493–520.

Audretsch, D. B., Thurik, A. R., Verheul, I., & Wennekers, S. (Eds.). (2002). *Entrepreneurship: Determinants and Policy in a European US Comparison.* Boston and Dordrecht: Kluwer Academic Publishers.

Bandura, A. (1986). *Social Foundations of Thought and Action: A Social Cognitive Theory.* Englewood Cliffs, NJ: Prentice Hall.

Bandura, A. (1997). *Self-efficacy: The Exercise of Control.* New York: Freeman.

Bandura, A. (2006). Guide for Constructing Self-Efficacy Scales. In F. Pajares & T. Urdan (Eds.), *Self-efficacy Beliefs of Adolescents* (Vol. 5, pp. 307–337). Greenwich, CT: Information Age Publishing.

Baron, R. A., Mueller, B. A., & Wolfe, M. T. (2016). Self-efficacy and Entrepreneurs' Adoption of Unattainable Goals: The Restraining Effects of Self-control. *Journal of Business Venturing, 31*(1), 55–71.

Bendassolli, P. F., Borges-Andrade, J. E., Gondim, S. M. G., & Makhamid, Y. M. (2016). Performance, Self-regulation, and Competencies of Entrepreneurs in Brazilian Creative Industries. *Psicologia: Teoriae Pesquisa* (Brasília. Online), *32*, 1–9.

Brockner, J., Higgins, E. T., & Low, M. B. (2004). Regulatory Focus Theory and the Entrepreneurial Process. *Journal of Business Venturing, 19*(2), 203–220.

Brown, J. M., Miller, W. R., & Lawendowski, L. A. (1999). The Self-regulation Questionnaire. In L. Vandecreek & T. L. Jackson (Eds.), *Innovations in Clinical Practice A Sourcebook* (pp. 281–292). Sarasota, FL: Professional Resource Press/Professional Resource Exchange.

Bryant, P. (2006). Improving Entrepreneurial Education Through Self-regulatory Skills. In *The NCIIA*, pp. 279–289.

Bryant, P. (2007). Self-regulation and Decision Heuristics in Entrepreneurial Opportunity Evaluation and Exploitation. *Management Decision, 45*(4), 732–748.

Bryant, P. T. (2009). The Self-regulatory Foundations of Entrepreneurial Ambidexterity (Summary). *Frontiers of Entrepreneurship Research, 29*(6), 15–37.

Byrne, B. M. (2010). *Structural Equation Modeling with AMOS* (2nd ed.). New York: Routledge.

Campo, J. L. M. (2011). Analysis of the Influence of Self-efficacy on Entrepreneurial Intentions. *Prospectiva, 9*(2), 14–21.

Cope, J. (2003). Entrepreneurial Learning and Critical Reflection: Discontinuous Events as Triggers for 'Higher-Level' Learning. *Management Learning, 34*(4), 429–450.

Forgas, J., Baumeister, R., & Tice, D. (2009). Psychology of Self-regulation: An Introductory Review. In J. Forgas, R. F. Baumeister, & D. Tice (Eds.), *Psychology of Self-regulation Cognitive, Affective, and Motivational Processes* (pp. 1–17). New York: Psychology Press.

Fornell, C., & Larcker, D. F. (1981). Evaluating Structural Equation Models with Unobservable Variables and Measurement Error. *Journal of Marketing Research, 18*(1), 39–50.

Frese, M. (2009). Toward a Psychology of Entrepreneurship—An Action Theory Perspective. *Foundations and Trends in Entrepreneurship, 5*(6), 435–494.

Ghonsooly, B., & Ghanizadeh, A. (2013). Self-Efficacy and Self-Regulation and Their Relationship: A Study of Iranian EFL Teachers. *The Language Learning Journal, 41*(1), 68–84.

Gibb, A. A. (2008). Entrepreneurship and Enterprise Education in Schools and Colleges: Insights from UK Practice. *International Journal of Entrepreneurship Education, 6*(2), 48.

Hair, G., Black, B., Babin, B., Anderson, R., & Tatham, R. (2010). *Multivariate Data Analysis* (7th ed.). Upper Saddle River, NJ: Pearson.

Hmieleski, K. M., & Baron, R. A. (2008). When Does Entrepreneurial Self-efficacy Enhance Versus Reduce Firm Performance? *Strategic Entrepreneurship Journal, 2*, 57–72.

Hu, L., Gu, J., Wu, J., & Lado, A. A. (2017). Regulatory Focus, Environmental Turbulence, and Entrepreneur Improvisation. *International Entrepreneurship and Management Journal, 14*, 129–148.

ILO. (2009). Global Employment Trends for Women International Labour Office—Geneva: ILO. Retrieved March 30, 2018, from http://www.ilo.org/wcmsp5/groups/public/@dgreports/@dcomm/documents/publication/wcms_103456.pdf

Jayadatta, S. (2017). Major Challenges and Problems of Rural Entrepreneurship in India. *IOSR Journal of Business and Management, 19*(9), 35–44.

Jiménez-Jiménez, D., & Sanz-Valle, R. (2011). Innovation, Organizational Learning, and Performance. *Journal of Business Research, 64*(4), 408–417.

Kuratko, D. F., Ireland, R. D., Covin, J. G., & Hornsby, J. S. (2005). A Model of Middle-Level Managers' Entrepreneurial Behavior. *Entrepreneurship Theory and Practice, 29*(6), 699–716.

Latha, K. L., Madhavaiah, C., & Murthya, B. E. V. V. N. (2008). Small Scale Entrepreneurship in India. *Serbian Journal of Management, 3*(2), 171–187.

Ling, H., & Venesaar, U. (2015). Enhancing Entrepreneurship Education of Engineering Students for Increasing Their Metacognitive Abilities: Analysis of Students' Self-assessments. *Engineering Economics, 26*(3), 333–342.

Lüthje, C., & Franke, N. (2003). The Making of an Entrepreneur Testing a Model of Entrepreneurial Intent Among Engineering Students at MIT. *R & D Management, 33*(2), 135–147.

Meyer, G. D. (2009). Commentary: On the Integration of Strategic Management and Entrepreneurship: Views of a Contrarian. *Entrepreneurship, Theory and Practice, 333*(1), 41–351.

Mezo, P. (2008). The Self-control and Self-management Scale (SCMS). *Journal of Psychopathology and Behavioral Assessment, 31*, 83–93.

Onoda, S. (2014). Examining the Relationships Between Self-efficacy, Effort Regulation Strategy Use, and English Vocabulary Skills. *Studies in Self-access Learning Journal, 5*(4), 357–371.

Patel, B., & Chavda, K. (2013). Rural Entrepreneurship in India: Challenge and Problems. *International Journal of Advance Research in Computer Science and Management Studies, 1*(2), 28–37.

Paulsen, M. B., & Gentry, J. A. (1995). Motivation, Learning Strategies and Academic Performance: A Study of College Finance Classroom. *Financial Practice and Education, 5*(1), 78–90.

Pihie, Z. A. L., & Bagheri, A. (2013). Self-efficacy and Entrepreneurial Intention: The Mediation Effect of Self-Regulation. *Vocations and Learning, 6*(3), 385–401.

Pittaway, L. A., & Cope, J. P. (2007). Entrepreneurship Education: A Systematic Review of the Evidence. *International Small Business Journal, 25*(5), 479–510.

Preacher, K. J., & Hayes, A. F. (2004). SPSS and SAS Procedures for Estimating Indirect Effects in Simple Mediation Models. *Behavior Research Methods, Instruments, and Computers, 36*, 717–731.

Preacher, K. J., Rucker, D. D., & Hayes, A. F. (2007). Addressing Moderated Mediation Hypotheses: Theory, Methods, and Prescriptions. *Multivariate Behavioral Research, 42*, 185–227.

Salkowitz, R. (2010). *Young World Rising: How Youth, Technology, and Entrepreneurship Are Changing the World from the Bottom Up*. Hoboken, NJ: Wiley.

Santhi, N., & Kumar, R. S. (2011). Entrepreneurship Challenges and Opportunities in India. *Bonfring International Journal of Industrial Engineering and Medical Science, 1*, 14–16.

Schunk, D. H. (1996). *Self-efficacy for Learning and Performance*. Paper presented at the meeting of the American Educational Research Association.

Schunk, D. H., & Zimmerman, B. J. (2007). Influencing Children's Self-efficacy and Self-regulation of Reading and Writing Through Modeling. *Reading & Writing Quarterly, 23*(1), 7–25.

Schwarzer, R., & Jerusalem, M. (1995). Generalized Self-efficacy Scale. In J. Weinman, S. Wright, & M. Johnston (Eds.), *Measures in Health Psychology: A User's Portfolio.* Causal and Control Beliefs (pp. 35–37). Windsor: NFER-NELSON.

Selladurai, M., & Priyadharshini, J. (2015). Opportunities and Challenges for Rural Entrepreneurship in India. *International Journal of Multidisciplinary Educational Research, 4*(5), 74–81.

Sobel, M. E. (1982). Asymptotic Confidence Intervals for Indirect Effects in Structural Equation Models. In S. Leinhardt (Ed.), *Sociological Methodology* (pp. 290–312). Washington, DC: American Sociological Association.

Swath, T., & Rao, K. V. (2013). Entrepreneurship in India. *International Journal of Social Science & Interdisciplinary Research, 2*(7), 104–116.

Trevelyan, R. (2011). Self-regulation and Effort in Entrepreneurial Tasks. *International Journal of Entrepreneurial Behavior & Research, 17*(1), 39–63.

Tseng, C. (2012). Linking Entrepreneurial Infrastructures and New Business Development: A Study of Entrepreneurship Development in Taiwan. *The Journal of Entrepreneurship, 21*(1), 117–132.

Tumasjan, A., & Braun, R. (2012). In the Eye of the Beholder: How Regulatory Focus and Self-efficacy Interact in Influencing Opportunity Recognition. *Journal of Business Venturing, 27*(6), 622–636.

Vedanthadesikan, G., & Pathmanathan, P. (2016). Rural Entrepreneurship–An Indian Scenario. *International Journal of Humanities & Social Science Studies, 3*(2), 88–95.

Venesaar, U., Ling, H., & Voolaid, K. (2011). Evaluation of the Entrepreneurship Education Programme in University: A New Approach. *The Amfiteatru Economic Journal, Academy of Economic Studies, Bucharest, Romania, 8*(30), 377–391.

Vinesh, P. (2014). Role of Women Entrepreneurs in India. *Global Journal of Finance and Management, 6*(5), 473–480.

CHAPTER 11

Assessing the Impact of Human Resource Management Practices on Teachers' Performance through HR Analytics

Jeevan Jyoti, Poonam Sharma, and Asha Rani

1 Introduction

In today's era, education is a crucial tool for everyone who wants to succeed in life. It enhances knowledge and vision of an individual and helps to cope with the diverse challenges of life. Further, it is difficult to meet one's desires and wants without education, as it helps to open various doors to the opportunities for achieving better prospects in life and upgrading one's career growth. In other words, it makes an individual strong mentally, socially and intellectually, by enhancing his/her knowledge level and technical skills that facilitate one to achieve a good position at his/her workplace.

In this context, the success of an education system depends upon the effort and contribution of the teachers. Teachers help to develop, preserve

J. Jyoti • A. Rani
Department of Commerce, University of Jammu, Jammu, Jammu and Kashmir, India

P. Sharma (✉)
Jammu and Kashmir State Board of School Education, Jammu, Jammu and Kashmir, India

H. Chahal et al. (eds.), *Sustainable Business Practices for Rural Development*, https://doi.org/10.1007/978-981-13-9298-6_11

and transmit knowledge among students (Coats 2000). They direct the student to perform better. They not only enhance the theoretical knowledge of the students but also help them to enhance their practical and technical skills. In this same vein, qualitative teaching increases students' chances of success (Bellows and Perry 2005).

People in the rural areas face many challenges, such as poverty and lack of awareness. Currently, the majority of places across rural India lack basic infrastructure and there is a need to ramp it up so that students have access to educational facilities. For the schools in rural areas, it is a massive challenge to find ways to motivate children to come to school regularly and take interest in academics. Such children depend on a single source, that is, teachers, who put their best efforts and time in nurturing them. A teacher also helps to develop the personality and promote cultural values among the students. Earlier research revealed that students who get proper teaching assistance from their teachers have a greater chance of securing good positions in academics as well as have better career advancement potential in future (Bellows and Perry 2005; Brown et al. 1999). Apart from this, it is imperative to improve the education sector by providing appropriate training and career development programmes to the teachers. Along with this compensation system, performance management of the teaching workforce needs to be improved. Although it is evident from the low allocation of budget for the education sector that the government of Jammu and Kashmir has been ignoring this sector (*India Today* 2019). Further, to improve the educational standard, one has to focus upon quality-based teaching. In this regard, performance management plays a crucial role to achieve the goals and objectives of higher quality educational standards. Teachers are considered as an important asset as they play a vital role in achieving the organizational performance. Though a lot of research has been conducted on the relationship between human resource management practices and organizational performance (Bibi et al. 2012; Khera 2010; Mondy 2010). Furthermore, a few researchers have investigated the impact of human resource practices (Bibi et al. 2012; Shaheen et al. 2013), like training and development (Khan et al. 2011), performance management, performance appraisal and empowerment on the university teachers' performance. Hence, rural areas have been selected to evaluate the impact of human resource management practices on teachers' performance. The same is presented in Fig. 11.1.

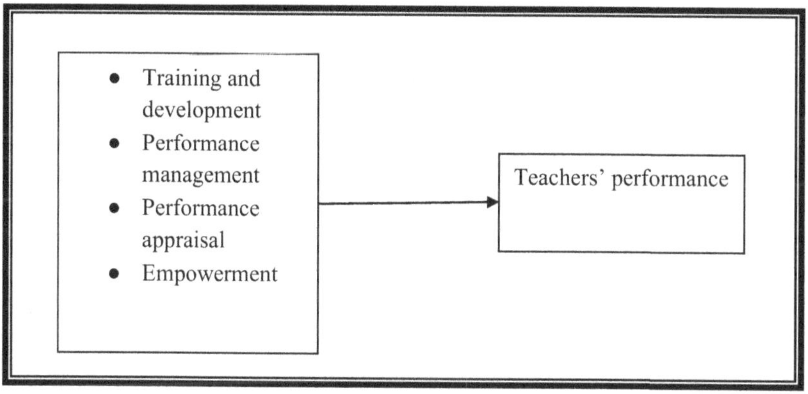

Fig. 11.1 Theoretical framework. (Source: Authors)

2 Theoretical Backgrounds and Hypotheses Development

Human resource management practices aim at enhancing the capabilities of the employees through proper training and development (Beltrán-Martín and Bou-Llusar 2018). These practices leverage organizational infrastructure to achieve organizational goals (Kellner et al. 2016). Chahal et al. (2016) viewed that human resource practices increase the knowledge and vision of the employees to achieve sustainable competitive advantage. On the same lines, Jyoti and Rani (2017) proved that high performance work system increases the efficiency of the employees that in turn improves the organizational performance, namely, financial performance, employee performance and operational performance. Besides this, human resource practices, such as training and development, performance-based compensation, performance appraisal and empowerment play a crucial role towards the performance of the employees (Hassan et al. 2013). In a nutshell, human resource practices enhance the employees' vision, skills and capabilities through ability-enhancing practices. It also motivates the employees by providing continuous feedback and performance-based compensation in order to improve the productivity of the organization (Chahal et al. 2016). Besides this, they give the opportunity to the employees to participate in the decision-making process that facilitates to extend innovative ideas, in turn to better the competitive position of the organization. Human resource management (HRM) comprises policies, practices and systems that influence employees' attitudes, behaviour and performance through learning and innovation (Noe et al. 2007). HR practices include human resource planning, recruitment, selection, screening, orientation, training, performance-based compensation, performance-appraisal, performance management, empowerment, competency development, job analysis, recognition and health and safety (Absar et al. 2010; Evans and Davis 2005; Mohamad et al. 2009; Osman et al. 2011; Stumpfy et al. 2010). Based on the existing literature, the following human resource management practices have been identified in the education sector and hence selected for the present study.

1. Training and development
2. Performance management
3. Performance appraisal
4. Empowerment

Training and Development and Teachers' Performance

Human resource practices, such as training and development, help to increase the knowledge and vision of the employees that, in turn, increases the productivity of an organization (Hong et al. 2012). Further, diverse goals of the organization can be achieved through effective training by motivating employees, enhancing their skills, knowledge and preparing them for future opportunities (Drummond1990). For example, extensive training facilities enhance the skill and knowledge of the employees (Birdi et al. 2008) that upgrade organizational performance. Further, such initiatives provide opportunity to the employees to share their know-how and experience with each other that helps to increase the productivity of the organization (Chen and Huang 2009). Thus, the following hypothesis has been framed.

H1 : Training positively affects teachers' performance

Performance Management and Teachers' Performance

Performance management refers to a systematic process for creating a common understanding among the employees, that is, what, when and how they can plan and achieve their goal. Planning includes the guidelines, action plan and strategies. It helps to manage the performance of the employees by proper guidance and feedback (Kagioglou et al. 2001; Storey 2002). Performance management helps to sustain and direct employees to work efficiently as well as effectively according to organizational needs that increase the performance of the organization (Armstrong and Baron 1998). Hence, the following hypothesis has been designed.

H2 : Performance management significantly affects teachers' performance

Performance Appraisal and Teachers' Performance

Performance appraisal is the continuous process of analysing and controlling the performance of the employees that helps to improve the efficiency of the organization (Carrel et al. 1995). To achieve maximum result in the education sector, various reforms in the form of implementation of human resource practices, like the performance appraisal system, have been undertaken. Performance appraisal is a continuous process that provides feedback to the employees in order to improve their performance. This system also helps the employees to know their strengths and weaknesses. Wati (2011) revealed that the implementation of human resource practices improves the performance of the teachers. Further, continuous and adequate feedback also helps to improve the teachers' performance (Odunlami 2014). Based on the earlier discussion, the following hypothesis has been framed.

H3 : Performance appraisal significantly enhances teachers' performance

Empowerment and Teacher Performance

Empowerment gives authority to an individual to think, behave and take actions; control work; and decision-making in an autonomous way (Gurbuz 2009). It can be divided into two instruments, namely, motivational and cognitive (Nykodym et al. 1994). Motivation is directly affected by trust, greater control of the work and setting of a higher goal. Cognitive instruments include upward communication and utilization of that transmitted information. Further, empowerment motivates the employees to participate in decision making and give suggestions for the welfare of the organization (Gurbuz 2009). Besides this, Hassan et al. (2013) demonstrated that human resource practices, like training, performance appraisal and empowerment, play a vital role in increasing the employees' commitment, loyalty and performance. In other words, empowerment also motivates the individuals to use the available knowledge and vision in the organization (Jyoti and Rani 2019) for better results. Hence, the following hypothesis has been designed.

H4 : Empowerment significantly affects teachers' performance

3 Methodology

Measures

A five-point Likert scale has been used in the present study ranging from strongly disagree (1) to strongly agree (5). Four HR practices have been used in the present study, namely, performance management (sample item: 'the information given to you about your job and the ways in which a position can be reached') (Conway and Monks 2008), performance appraisal (sample item: 'you are appraised by schools head/principle') (Chen and Huang 2009), training (sample item: 'your department encourages teachers to undertake continuous training') (Waseem et al. 2013; Wei et al. 2010) and empowerment (sample item: 'teacher empowerment is highly valued in the school') (Pare and Tremblay 2000). Teacher performance scale has been developed on the lines of Gusthart et al. (1997) and Waseem et al. (2013). The sample item is 'the teachers are getting competitive salaries/benefits'.

Data Collection

Data has been collected from the teachers of rural private school of Rajouri district of Jammu and Kashmir. The total population is 607. All the teachers have been contacted for data collection on convenience bases. Out of 607, only 245 teachers gave the required responses. The response rate is 41 per cent.

4 Results

Exploratory Factor Analysis

Factor analysis has been conducted to summarize the data and identify the factor (Hair et al. 2010). The scale of HR practices originally consisted of 25 statements. After applying EFA, these 25 items got reduced to 21, which are clubbed under four factors. Similarly, teacher's performance scale initially consisted of six statements that got reduced to five under single factor. The EFA results are presented in Table 11.1.

Factor Confirmation

Zero-order models have been designed for training and development, performance management, performance appraisal, empowerment and teacher performance. Figures 11.2, 11.3, 11.4, 11.5 and 11.6 display all the goodness of fit (greater than 0.90) and badness of fit (less than 0.08) all criteria are within the prescribed limits, which established that all the models have excellent confirmation to be used for further analysis.

Reliability and Validity Analysis

All the constructs have been thoroughly investigated for the reliability and validity. High Cronbach's alpha value (greater than 0.70) and composite reliability values (greater than 0.70) proved that the data was quite reliable to be used. Convergent and discriminant validity have also been proved as high standardized estimates and average variance explained (greater than 0.5) proved the convergent validity (Table 11.2). Further, comparison of AVE with squared correlations provided the support for discriminant validity (Table 11.3).

Hypotheses Testing

In order to check various hypothesized relationships, structural equation modelling (SEM) has been used in the present research. The relationships between training and development, performance management, performance appraisal, empowerment and teacher performance have been assessed through SEM. The SRW (standard regression weight) came out to be 0.39 ($p < 0.001$), 0.45 ($p < 0.001$), 0.31 ($p < 0.001$) and 0.31

Table 11.1 Summary of result of exploratory factor analysis

Factor	*Mean*	*SD*	*FL*	*C*	*E.V*	*V.E (%)*	*KMO*
HRM Practices	*4.04*					*70.467*	*0.899*
Training & Development	*4.01*	*0.61*			*2.76*	*8.627*	*0.858*
T&D1	4.12	0.78	0.659	0.624			
T&D 2	4.02	0.81	0.689	0.62			
T&D 3	3.9	0.83	0.68	0.584			
T&D 4	4.03	0.77	0.589	0.511			
Performance Management	*4.07*	*0.55*			*10.652*	*33.286*	
PM1	4.22	0.78	0.539	0.593			
PM2	4.02	0.71	0.535	0.548			
PM3	4.14	0.81	0.52	0.518			
PM4	3.97	0.83	0.751	0.657			
PM5	4.13	0.77	0.667	0.527			
PM6	3.94	0.92	0.63	0.521			
PM7	4.11	0.76	0.643	0.624			
Performance Appraisal	*4.08*	*0.53*			*3.28*	*10.25*	
PA1	3.94	0.83	0.734	0.548			
PA2	4.11	0.84	0.713	0.672			
PA3	4.22	0.71	0.67	0.59			
PA4	4.08	0.79	0.557	0.557			
Empowerment	*4.01*	*0.54*			*1.219*	*3.808*	
EMP1	3.97	0.88	0.585	0.517			
EMP2	4.18	0.68	0.51	0.583			
EMP3	4.07	0.73	0.675	0.601			
EMP4	3.94	0.81	0.534	0.532			
EMP5	3.87	0.82	0.564	0.619			
EMP6	4.04	0.79	0.639	0.545			
Teacher Performance	*3.27*	*0.57*			*5.525*	*60.824*	*0.858*
TP1	3.89	0.81	0.541	0.602			
TP2	2.22	0.90	0.730	0.532			
TP3	3.9	0.77	0.565	0.589			
TP4	2.29	0.87	0.525	0.501			
TP6	4.03	0.84	0.646	0.623			

Source: Authors

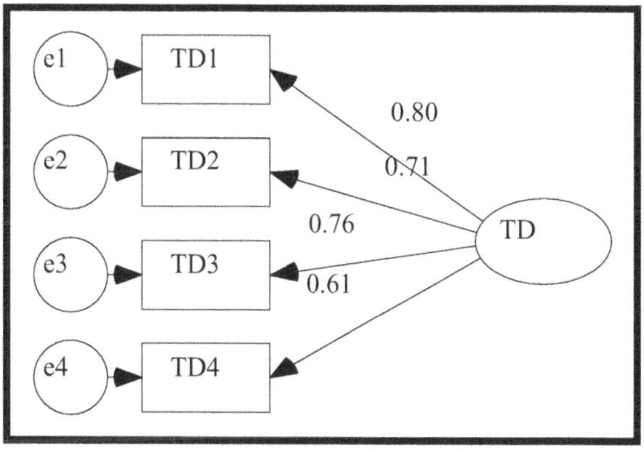

Fig. 11.2 Measurement model of training and development

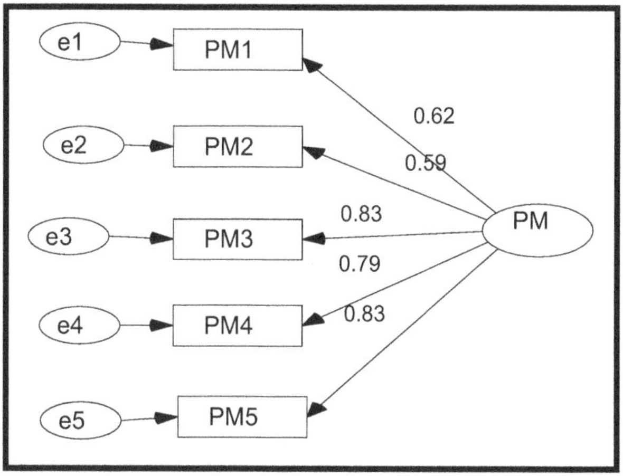

Fig. 11.3 Measurement model of performance management

Fig. 11.4 Measurement model of performance appraisal

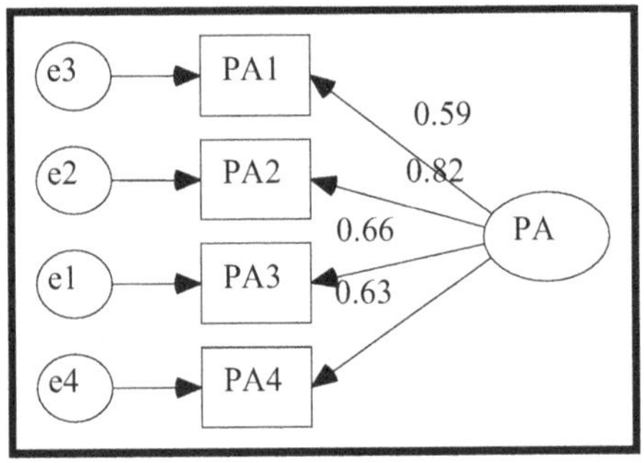

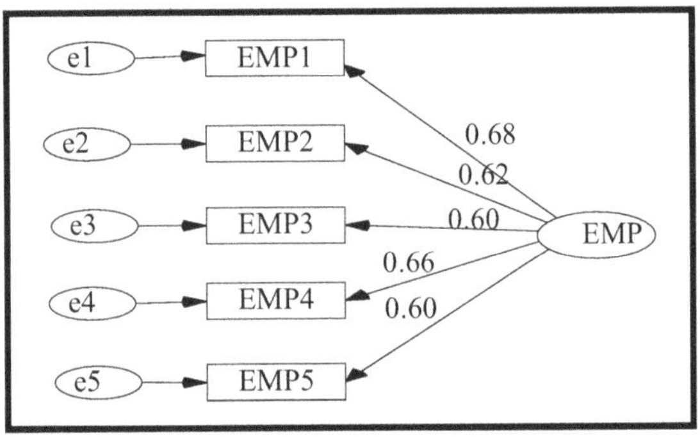

Fig. 11.5 Measurement model of empowerment

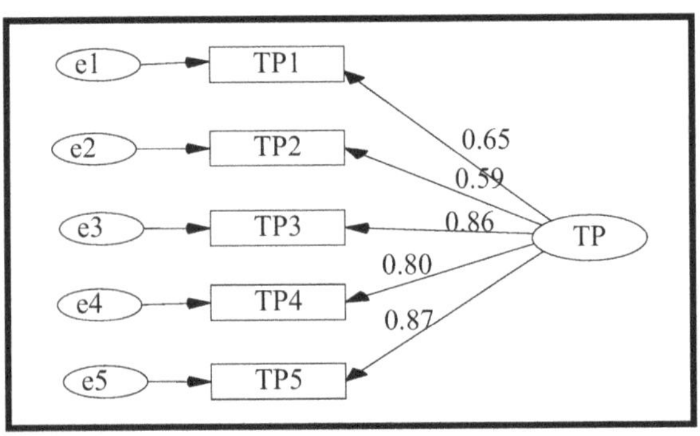

Fig. 11.6 Measurement model of teachers' performance

Table 11.2 Reliability and validity analysis

Constructs	*Dimensions*	*AVE*	*C.R*	*Alpha Value*
HRM Practices		0.863	0.941	0.919
	T&D	0.684	0.848	0.778
	PM	0.776	0.874	0.822
	PA	0.673	0.847	0.748
	EMP	0.817	0.858	0.786
Teachers' Performance		0.627	0.853	0.793

Source: Authors

Table 11.3 Discriminant validity and correlation analysis

Constructs	*Correlation matrix*					
	HRM practices	*T&D*	*PM*	*PA*	*EMP*	*TP*
HRM practices	*0.863*					
T&D	(0.625) 0.791**	*0.684*				
PM	(0.484) 0.696**	(0.521) 0.722**	*0.776*			
PA	(0.386) 0.622**	(0.417) 0.646**	(0.358) 0.599**	*0.673*		
EMP	(0.496) 0.704**	(0.524) 0.724**	(0.472) 0.687**	(0.383) 0.619**	*0.817*	
TP	(0.352) 0.593**	(0.376) 0.613**	(0.361) 0.601**	(0.345) 0.587**	(0.283) 0.532**	*0.627*

Source: Authors

Note: Values on the diagonal axis represent average variance extracted and values in parenthesis represent squared correlation between the constructs. The values with asterisk represent correlations values

($p < 0.001$), respectively, which revealed that training and development, performance management, performance appraisal and empowerment significantly and positively affect teachers' performance (Figs. 11.7, 11.8, 11.9 and 11.10). So, all hypotheses stand accepted.

Further, integrated model revealed that HRM practices positively affect teacher performance (Fig. 11.11). Model fit indices are also good (χ^2/df = 1.785, RMR = 0.047, GFI = 0.886, AGFI = 0.861, CFI = 0.908, RMSEA = 0.055).

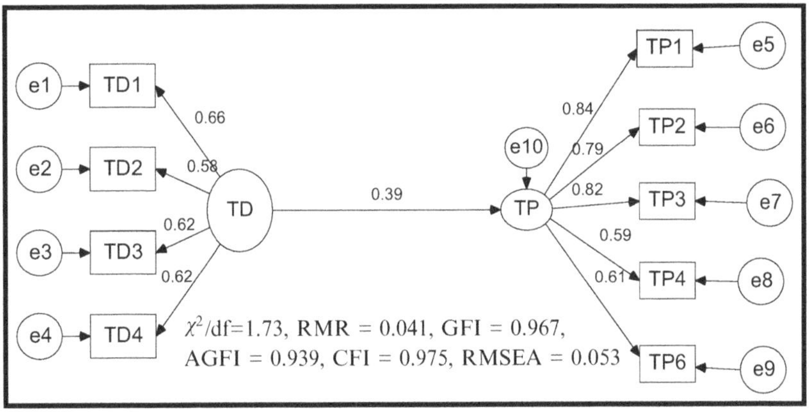

Fig. 11.7 Impact of training and development on teachers' performance

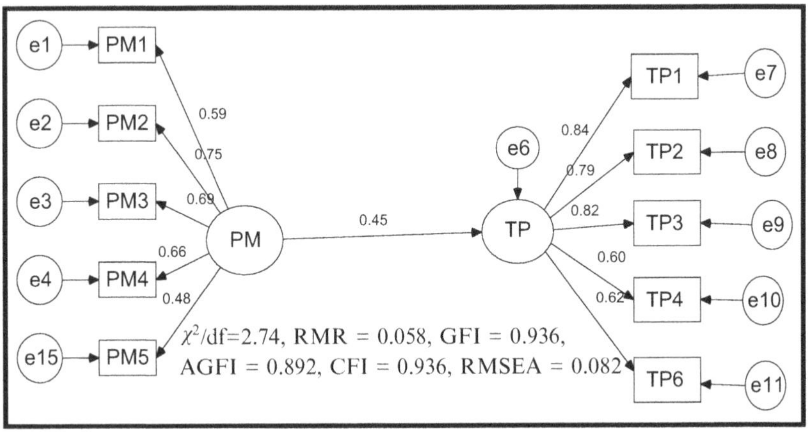

Fig. 11.8 Impact of performance management on teachers' performance

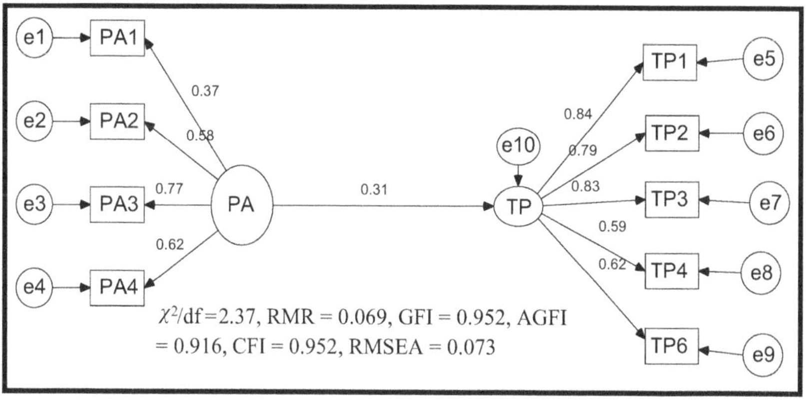

Fig. 11.9 Impact of performance appraisal on teachers' performance

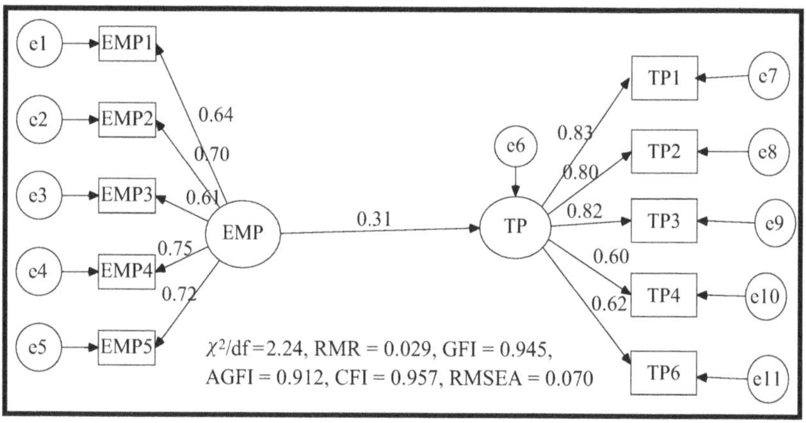

Fig. 11.10 Impact of empowerment on teacher performance

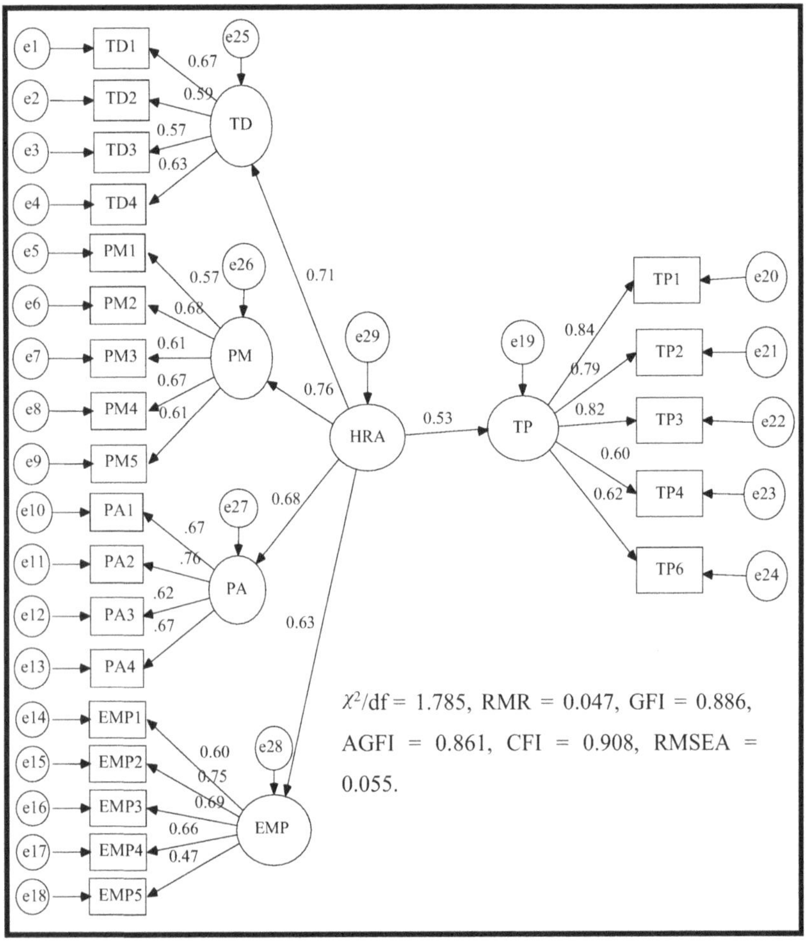

Fig. 11.11 Integrated model. Key: TD = training and development, PM = performance management, PA = performance appraisal, EMP = empowerment, TP = teachers' performance, e1–e19 = error term of manifest variables

5 Discussion

This chapter empirically investigates the impact of human resource management on teacher performance through HR analytics in rural schools of Jammu and Kashmir. Result of the present research revealed that all HR practices are significant for teachers' performance, which is in line with Shaheen et al. (2013). Results indicated that training and development boosts up the skill, knowledge and competence of the teachers that improves performance and motivates the teachers to give their best. It also provides the platform for the teachers to carry out their role and responsibilities optimally, which matches with the predetermined standard. Through good performance appraisal a teacher gets feedback, set performance goals, which ultimately improves their performance. Performance management helps teachers to make full use of their skills and abilities. Teachers are involved in decision making related to school issues. Further, teachers' empowerment is highly valuable in schools, which encourages teachers to perform better.

6 Implications

The findings of this study have both theoretical and managerial implications for teachers and researchers to successfully promote human resource practices and to establish theoretical underpinning of the conceptual framework. These implications have been discussed further.

Theoretical Implications

This study contributes to the existing knowledge of human resource management practices literature by empirically modelling the relationships in education sector. This investigation gives a roadway for further investigation. We empirically proved the role of training and development, performance management, performance appraisal on teachers' performance in rural India of Jammu and Kashmir.

Managerial Implications

The finding suggested that there are inadequate training programmes. In this context, the management should organize training programmes and make them compulsory for newly-recruited teachers. These training courses will act as add-on knowledge about the latest changes and advancements as well as practical knowledge among the existing teachers. Besides this, latest methods and mode of teaching should be communicated by the concerned authorities, so that they can upgrade their performance. This will ultimately encourage them to accept challenges and prepare them to cope with the changes. In order to get the best results, the management should recognize the best performer by rewarding those teachers, who performed their job well. The management should also organize reunions to honour the best achievers. Further, management should not only welcome senior teachers' participation but also encourage junior teachers to participate in the policy-making and educational issues.

7 Limitation and Future Research

No doubt the present chapter has put forward many theoretical and managerial implications, but some of the limitations have been identified. First, the cross-sectional nature can create the problem of common method bias. Second, only one outcome has been examined. Future researches can examine students' academic performance as well as school's performance too.

References

Absar, M. N., Nimalathasan, B., & Jilani, M. A. K. (2010). Impact of HR Practices on Organisational Performance in Bangladesh. *International Journal of Business and Information Technology, 3*(2), 15–19.

Armstrong, M., & Baron, A. (1998). *Performance Management: The New Realities.* London: Institute of Personnel and Development.

Bellows, L. H., &Perry, A.2005. *Assessing Graduate Student Mentorship and Development.* Unpublished manuscript. Retrieved January 23, 2015, from www.unl.edu.

Beltrán-Martín, I., & Bou-Llusar, J. C. (2018). Examining the Intermediate Role of Employee Abilities, Motivation and Opportunities to Participate in the Relationship between HR Bundles and Employee Performance. *BRQ Business Research Quarterly, 21*(2), 99–110.

Bibi, A., Lanrong, Y., Haseeb, M., & Ahmad, I. (2012). The Effect of Human Resource Management Practices on Employees' Job Satisfaction in the Universities of Pakistan. *Business Management Dynamics, 1*(12), 1–14.

Birdi, K., Clegg, C., Patterson, M., Robinson, A., Stride, C., Wall, T., & Wood, S. (2008). The Impact of Human Resource and Operational Management

Practices on Company Productivity: A Longitudinal Study. *Personnel Psychology, 61*(3), 467–501.

Brown, M. C., Davis, G. L., & McClendon, S. A. (1999). Mentoring Graduate Students of Color: Myths, Models and Modes. *Peabody Journal of Education, 74*(2), 105–118.

Carrel, M. R., Elbert, N. F., & Hatfield, R. D. (1995). *Human Resource Management: Global Strategies for Managing a Diverse Workforce*. New Jersey: Prentice Hall.

Chahal, H., Jyoti, J., & Rani, A. (2016). Bundled Effect of High Performance Human Resource Practices on Business Performance: Role of Organisational Learning. *Global Business Review, 17*(3(S)), 107S–132S.

Chen, C. J., & Huang, J. W. (2009). Strategic Human Resource Practices and Innovation Performance: The Mediating Role of Knowledge Management Capacity. *Journal of Business Research, 62*(1), 104–114.

Coats, L. T. (2000). Interpersonal Behaviour and the Community College Department Chairperson. *Community College Journal of Research and Practice, 24*(3), 773–783.

Conway, E., & Monks, K. (2008). HR Practices and Commitment to Change: AnEmployee-Level Analysis. *Human Resource Management Journal, 18*(1), 72–89.

Drummond, K. E. (1990). *Human Resource Management for the Hospitality Industry*. New York: Van Nostrand Publishing.

Evans, W. R., & Davis, W. D. (2005). High-Performance Work Systems and Organisational Performance: The Mediating Role of Internal Social Structure. *Journal of Management, 31*(5), 758–775.

Gurbuz, S. (2009). The Effect of High Performance HR Practices on Employees' Job Satisfaction. *Journal of the School of Business Administration, 38*(2), 110–123.

Gusthart, J. L., Kelly, I. M., & Rink, J. E. (1997). The Validity of the Qualitative Measures of Teaching Performance Scale as a Measure of Teacher Effectiveness. *Journal of Teaching in Physical Education, 16*(2), 196–210.

Hair, J. F., Black, W. C., Babin, B. J., Anderson, R. E., & Tatham, R. L. (2010). *Multivariate Data Analysis* (7th ed.). Upper Saddle River, NJ: Pearson Prentice Hall.

Hassan, M. U., Nawaz, M. S., Abbas, G., & Sajid, M. I. (2013). Impact of High Performance Work Practices on Employee Loyalty and Financial Performance through Mediation of Employee Satisfaction: An Empirical Evidence from the Financial Sector of Pakistan. *Middle-East Journal of Scientific Research, 16*(8), 1037–1046.

Hong, E. N. C., Hao, L. Z., Kumar, R., Ramendran, C., & Kadiresan, V. (2012). An Effectiveness of Human Resource Management Practices on Employee Retention in Institute of Higher Learning: A Regression Analysis. *International Journal of Business Research and Management, 3*(2), 60–79.

India Today. 2019. Budget 2019: There's Money for World Class Institutes, But Not Much for Education. Retrieved July 5, 2019, from https://www.indiatoday.in/budget-2019/story/budget-2019-world-class-institutes-education-sector-1562891-2019-07-05.

Jyoti, J., & Rani, A. (2017). High Performance Work System and Organisational Performance: Role of Knowledge Management. *Personnel Review, 46*(8), 1770–1795.

Jyoti, J., & Rani, A. (2019). Role of Burnout and Mentoring between High Performance Work System and Intention to Leave: Moderated Mediation Model. *Journal of Business Research, 98*, 166–176. https://doi.org/10.1016/j.jbusres.2018.12.068.

Kagioglou, M., Cooper, R., & Aouad, G. (2001). Performance Management in Construction: A Conceptual Framework. *Construction Management and Economics, 19*(1), 85–95.

Kellner, A., Townsend, K., Wilkinson, A., Greenfield, D., & Lawrence, S. (2016). The Message and the Messenger: Identifying and Communicating a High Performance "HRM Philosophy". *Personnel Review, 45*(6), 1240–1258.

Khan, R. A. G., Khan, F. A., & Khan, M. A. (2011). Impact of Training and Development on Organizational Performance. *Global Journal of Management and Business Research, 11*(7), 1–7.

Khera, S. N. (2010). Human Resource Practices and Their Impact on Employee Productivity: A Perceptual Analysis of Private, Public and Foreign Bank Employees in India. *DSM Business Review, 2*(1), 65–86.

Mohamad, A. A., Lo, M. C., & La, M. K. (2009). Human Resource Practices and Organisational Performance. Incentive as a Moderator. *Journal of Academic Research in Economics, 1*(2), 229–242.

Mondy, R. W. (2010). *Human Resource Management* (11th ed.). Upper Saddle River, NJ: Pearson.

Noe, R. A., Hollenbeck, J. R., Gerhart, B., & Wright, P. M. (2007). *Human Resource Management: Gaining a Competitive Advantage*. New York: McGraw-Hill.

Nykodym, N., Simonetti, J. L., Nielsen, W. R., & Welling, B. (1994). Employee Empowerment. *Empowerment in Organisations, 2*(3), 45–55.

Odunlami, I. B. (2014). Impact of Customer Satisfaction and Customer Retention on Customer Loyalty: A Case Study of Enterprise Bank in Oyo. *International Journal of Education and Research, 2*(9), 427–450.

Osman, I., Ho, C. F., & Galang, M. C. (2011). 'The Relationship between Human Resource Practices and Firm Performance: An Empirical Assessment of Firms in Malaysia. *Journal of Business Strategy Series, 12*(1), 41–48.

Pare, G., &Tremblay, M.2000. The Measurement and Antecedents of Turnover Intentions among IT Professionals. Retrieved November 3, 2017, fromhttp://www.cirano.umontreal.ca/pdf/publication/2000s-33.pdf.

Shaheen, I., Sajid, M. A., & Batool, Q. (2013). Impact of HR Practices on Performance of Teachers in Colleges of Azad Kashmir. *International Journal of Management Sciences & Business Research, 2*(3), 13–21.

Storey, A. (2002). Performance Management in Schools: Could the Balanced Scorecard Help? *School Leadership & Management, 22*(3), 321–338.

Stumpfy, S. A., Doh, J. P., Tymon, W. G., & J.R. (2010). The Strength of HR Practices in India and Their Effects on Employee Career Success, Performance, and Potential. *Journal of Human Resource Management, 49*(3), 353–375.

Waseem, S. N., Faroogi, R., & Afshan, S. (2013). Impact of Human Resource Management Practices on Teachers' Performance: A Mediating Role of Monitoring Practices. *Journal of Education and Social Sciences, 1*(2), 31–35.

Wati, H. (2011). The Effectiveness of Indonesian English Teachers Training Programs in Improving Confidence and Motivation. *International Journal of Instruction, 4*(1), 79–104.

Wei, Y. C., Han, T. S., & Hsu, I. C. (2010). High-Performance HR Practices and OCB: A Cross-Level Investigation of a Casual Path. *International Journal of Human Resource Management, 21*(10), 1631–1648.

Index

H. Chahal et al. (eds.), *Sustainable Business Practices for Rural Development*, https://doi.org/10.1007/978-981-13-9298-6

The manufacturer's authorised representative in the EU is Springer Nature Customer Service Centre GmbH, Europaplatz 3, 69115 Heidelberg, Germany. If you have any concerns regarding our products, please contact ProductSafety@springernature.com

Printed and bound by CPI Group (UK) Ltd, Croydon, CR0 4YY
25/09/2026
02222897-0004